THE END OF A REVOLUTION:
THE LAST DAYS OF STATIONARY STEAM

Colin Bowden

Published by

Ashbourne Hall, Cokayne Ave
Ashbourne, Derbyshire DE6 1EJ England
Tel: (01335) 347349 Fax: (01335) 347303
e-mail: landmark@clara.net
web site: www.landmarkpublishing.co.uk

ISBN 13: 978-1-84306-352-0

British Library Cataloguing in Publication Data: a catalogue record for this book is available from the British Library.

Printed by Cromwell Press Ltd, Trowbridge
Designed by Michelle Prost
Edited by Ian Howe

Front cover: Torpoint, Ministry of Defence, Thanckes Oil Fuel Depot

Back cover: Fleet, Grab Dredger Perseverance, on Basingstoke Canal

THE END OF A REVOLUTION:
THE LAST DAYS OF STATIONARY STEAM

Colin Bowden

Landmark Publishing

Contents

Contents

Contents

Contents

Contents

Contents

Foreword

Still to be seen during the 1970s and 1980s were the last remaining examples of the use of the reciprocating steam engine as a source of industrial power in the UK. A few were quite spectacular: a 12,000 horsepower rolling mill engine in Sheffield (which worked for the last time in 1978 and is now in the Kelham Island Museum), and the 2,650 horsepower engine at the Ellenroad Ring Mill, Milnrow (which stopped in 1975 and is still preserved on site) were examples; and as a very small remnant of what once had been, there were still at least 40 textile mill engines at work in Lancashire and Yorkshire, and more than 50 collieries with steam winding engines.

By the 1990s, however, little of significance was left working. The last mill engines of any size had stopped: at Queen Street Mill, near Burnley, in 1982; the rolling mill engines at Thomas Walmsley's Atlas Forge, Bolton, in 1983; and the engine at William Barker's tannery, Otley, in 1988 (see No. 173 in this book). Other significant stoppages were of the last waterworks steam engine at Waddon Pumping Station, Croydon, in 1982; and that of the last coal winder at Sutton Manor Colliery, Lancashire, which ended its working life in 1991. Today, the number of working engines has been reduced to a doubtful handful. Even what were probably the last Belliss & Morcom high-speed enclosed engines (at the New Cheshire Salt Works, Wincham) came to a halt in June 2006.

The photographs selected for inclusion in this book are of a representative sample of the engines which survived into this final period. They are part of the product of a survey which, starting in 1972, aimed at recording steam engines in the UK remaining on their working sites. This was carried out jointly with the late George Cooper until 1991 and in more recent years in co-operation with fellow members of the International Stationary Steam Engine Society. It followed on from and overlapped slightly with the work of George Watkins. Active over a much longer and mainly earlier period (c.1930–1975), Watkins was able to record a very much greater number of engines, but inevitably he was unable to cover everything there was to see, and some of the less interesting and smaller examples were omitted.

This book, it is hoped, will provide a partial overview of industrial steam in the final years of the twentieth century, and also be a very modest supplement to the ten volumes of Watkins's Stationary *Steam Engines of Great Britain, 2000–2005*. So engines included in this and in his *Textile Mill Engine*, 2nd edn, 1999, are excluded. But in addition to a number of the larger engines missed from the Watkins volumes, it has been possible to include here more examples of the smaller, simpler engines – the "workhorses", built in the greatest numbers and used for an enormous variety of different purposes.

Of the 205 engine sites included, manufactures and activities represented in the largest numbers are:

Textile and clothing manufacture	26
Coal mines	22
Water & sewage works	20
Sawmills & furniture manufacture	15
Brickworks & potteries	13
Coke, gas & chemical works	12
Breweries	12

The remaining sites fall into seventeen or more different types of activity.

The changes which have occurred at the 205 sites during the period of the survey help to confirm that these were indeed the final years. In the 1970s, about 80 of the sites had engines which were working commercially. Today, perhaps five have engines which might be considered as seriously operative (see Nos. 41, 74, 107, 147 and 205), although No. 74 (Frogmore Mills) and No. 205 (Balloch Slipway) are linked to museum operations and in only semi-commercial use. In fact, there may now be no more than about half a dozen working engines, in all, in the UK – probably only the Hook Norton Brewery engine in Oxfordshire (see *Stationary Steam Engines in Great Britain*, Volume 6, No. 120) can be added to the five above. The industrial archaeology of the stationary steam engine, in the sense that it is concerned with the artefact in its context, is also disappearing. Only 75 of the sites illustrated still have in situ engines, and some of these have had their archaeological integrity degraded to a greater or lesser degree by preservation activity or other changes; 85 have had their engines removed for preservation, and 45 their engines scrapped.

The arrangement of the book is very similar to that of the George Watkins volumes, with the sites ordered geographically and alphabetically using the same historic county names and boundaries. Also, following the Watkins example, one or two non-stationary sites and an alternative power source have been included. As six-figure grid references were to hand, it seemed reasonable to include them.

There are, of course, considerable problems regarding the tabulated statistical information, particularly that relating to cylinder sizes and steaming conditions:

(1) Cylinder diameters in particular are well known for the variability of the figures which have been quoted. Even documentary sources such as engine builders' registers and indicator test reports, which have been used when available, are not completely reliable – for various reasons. Builders' catalogues are sometimes a useful guide but more often hearsay evidence or other secondary sources have to be used, or as a last resort approximations (of varying likelihood) based on external measurement. Only rarely is it possible actually to measure the cylinder bore.

(2) Steaming conditions (steam pressure, rpm and horsepower) also present difficulties, partly because there is usually no one set of correct figures – there are likely to be changes over the years or shorter term fluctuations. Where possible, observed working conditions (with hearsay or estimated horsepower figures) are used, although these may reflect reduced boiler pressures and/or loads rather than full-load working. Alternatively, in some cases indicator test records or makers' design figures have been used. There is additional ambiguity in the case of horse-power figures quoted simply as "hp" rather than more specifically as "ihp", "bhp" (or "whp"). Also one or two of them may be nominal figures ("nhp"), which are often related to cylinder diameter and stroke, and rather meaningless.

Cylinder and flywheel dimensions are given in imperial measurements, with the exception of those for a very small number of engines of Continental origin for which metric figures are given.

Sources and Acknowledgements

A very large number of sources have been used and it would take up too much space to include all the reference details. Most of the information comes from notes made at the time of the visits, including word-of-mouth from the drivers and engineers. Also used have been the visit notes made by George Watkins relating to engines which he visited, but which for one reason or another were not included in the volumes referred to above, but which are included in this book; and George Cooper's visit notes, also now held with his photographs as part of English Heritage's National Monuments Record, Swindon. In some cases, when available, engine builders' records have been used – this applies for example to Belliss & Morcom; Davey, Paxman; Richard Garrett; Hick, Hargreaves; Markham; Robey; Ruston, Proctor; Walker Brothers; and Worthington-Simpson.

(Some guidance to the locations of these records is provided by *Records of British Business and Industry 1760–1914. Metal Processing and Engineering*, HMSO, 1994.)

Much has been derived from secondary printed sources – contemporary or otherwise. These include numerous periodicals in particular *The Engineer* and *Engineering*; and the many contributions by various authors to the (International Stationary Steam Engine Society's) *ISSES Bulletin*. The important long-running series of articles by Geoff Hayes in *Engineering in Miniature* should also be mentioned (particularly that on the use of steam power in the textile industry, "Steam in Fashion", dating from the period 2001–2005).

Over the years, an enormous number of people have helped with information and advice, and to them - far too many to acknowledge individually – I owe heartfelt thanks. But I should mention specifically those who have read through and helped with the text and production of the book. Chris Allen has helped with suggestions and eliminated a number of errors (needless to say he cannot be held responsible for those which remain). My wife Sally has improved its English (and has for more years than I wish to think about lived with the excesses of steam engine mania in all its forms). Tony Woolrich has provided encouragement and guidance during the work's gestation and finalisation, which also owes much to the unusually difficult design work by Michelle Prost. In conclusion, above all, thanks are due to all those company owners and engineers who were, almost without exception, generous with information and enthusiastic in providing the opportunity to visit and photograph their engines.

Bibliographical Note

It is perhaps unnecessary to repeat the bibliography to be found in *Stationary Steam Engines of Great Britain Volume 1: Yorkshire*, 2000, but the most useful Watkins references, additional to those given above, should perhaps be noted:

The Stationary Steam Engine, 1968

The Industrial Archaeology of the Stationary Steam Engine, 1976 (with R.A. Buchanan)

The Steam Engine in Industry, 2 vols, 1978–9

The following in particular can also be recommended and may be found useful:

George Watkins and Frank Wightman, *Steam Power. An Illustrated Guide* (in *Industrial Archaeology* Vol. 4, No. 2, 1967), reprinted 1982. (This includes Frank Wightman's drawings of most of the basic engine types.)

Richard L. Hills, *Power from Steam: A History of the Stationary Steam Engine*, 1989. (Although not quite as comprehensive as the title might suggest, this is the only good modern history of the reciprocating steam engine in English.)

BEDFORDSHIRE

1. Stotfold, Fairfield Hospital (TL204352)

Photo taken/Reference: 20.2.1984/433-36
Type: Horizontal single cylinder
Maker and Date: Robey & Co. Ltd., Lincoln. No. 24462. 1905
Cylinder dimensions: 12in x 2ft. Drop valves; slide exhaust valves
Hp: 94 (max des) **Rpm:** 120 **Psi:** 115 **Flywheel diameter:** 8ft
Service: Driving pump, laundry machinery etc.

This is a Robey E Class engine, a design associated with John Richardson, delivered here new to what was then the Three Counties Asylum. All the valves were worked by eccentrics on the side shaft driven by skew gears, which replaced the bevel gearing used on earlier engines of this type. This shaft also served to drive the Richardson governor standing over the cylinder and controlling the inlet valve trip gear. The exhaust valves have their own eccentrics and are grid iron slide valves working at right-angles to and below the cylinder. Exhaust steam was presumably used for drying in the laundry. Drive was by canvas belt (9ins wide) from a 5ft diameter pulley to fast and loose pulleys on an overhead lineshaft, from which belts were taken to a 3-throw vertical well pump (still in situ), laundry machinery, cobblers' shop and tailors' sewing machines. The washing machines (replaced by then?) were not seen, and the cobblers' and tailors' shops had gone by 1984. The engine had been taken out of use, possibly in 1939, when it was said a new electric pump had been installed, and in 1986 it was removed to the Launceston Museum of Power, Cornwall.

BERKSHIRE

2. Arborfield, Bearwood College laundry (SU770687)

Photo taken/Reference: 24.5.1975/31-11
Type: Horizontal single cylinder
Maker and Date: Marshall Sons & Co. Ltd., Gainsborough. No. 32272. 1900
Cylinder dimensions: c.8ins x 1ft. Slide valve
Psi: 100?
Flywheel diameter: 4ft 6ins
Service: Belt drive to laundry machinery via lineshaft

The College laundry had clearly been out of use for some time when seen in 1975, but the engine and two rotary washing machines by R.G. Whitaker Ltd., Kingston-on-Thames remained. Also in situ was the small boiler by J. Hickey & Sons Ltd., Richmond – with a single furnace tube and return tubes on each side. This was an engine typical of those manufactured in huge numbers for small powers, with a looped frame supporting a bent crank between the two bearings, and overhung flywheel and belt pulley. A Pickering type governor was used for control of the steam throttle valve and the exhaust would presumably have been used for drying in the laundry. The engine was removed for private preservation in 1979. (Bear Wood House was built 1865–74 for John Walter III, proprietor of *The Times*, and he appears also to have established the laundry, but the house was abandoned for a period in the early 1900s. The engine was possibly installed second-hand, after 1919 when Bear Wood was sold to house the Merchant Seamen's Orphanage. It later became, and still remains, the Royal Merchant Navy School.)

1

3. Englefield, Englefield Estate Works sawmill (SU628723)

Photo taken/Reference: 25.10.1981/312-18
Type: Horizontal single cylinder
Maker and Date: Ransomes, Sims & Jefferies Ltd., Ipswich. No. 12885. 1899
Cylinder dimensions: 13in x 2ft. Slide valve; expansion slide valve
Hp: 100? (des. max) **Rpm:** 120? (des.) **Psi:** 100
Flywheel diameter: 8ft 2in
Service: Driving sawmill machinery by belts

The two engines at the Englefield Estate sawmill probably stopped working in 1951, but in the early 1970s were restored to working order by members of the Thames Valley Traction Engine Club. They are still kept in immaculate condition and steamed occasionally. The Ransomes, Sims & Jefferies engine was installed in January 1900 by the engineer, John Wilder of Reading, as a replacement for the original – another horizontal engine, which worked here from 1856–1899. The Cornish boiler, c.5ft 6in x 20ft, fed with wood waste, was put in at the same time. The engine, with very clean lines, girder frame, trunk guide and disc crank, has a variable expansion slide valve driven via Hartnell governor-controlled linkage. A third eccentric drives a diagonal boiler feed pump. There are three belt pulleys of varying diameters, one of which, next to the flywheel, drives to an overhead lineshaft. In a cupboard, next door, are the other two, from which a belt drove down to an underfloor shaft directly below. Some, at least, of the machinery survives, including a large horizontal saw by Kirchner of Leipzig.

4. Englefield, Englefield Estate Works sawmill (SU628723)

Photo taken/Reference: 25.10.1981/312-36
Type: Horizontal single cylinder
Maker and Date: Clayton, Shuttleworth & Co., Lincoln. No. 5197. 1863
Cylinder dimensions: c.8in x 14in. Slide valve
Hp: 8
Flywheel diameter: 6ft
Service: Driving by belt from 57in diameter pulley to sawbench by Worssam & Co., Chelsea

A much earlier engine than the Ransomes, with flat bed and small Watt governor as distinctive features, it can, luckily, be dated. The Serial No. 5197 and date appear on the circular brass maker's plate behind the steam inlet valve control handle, and its delivery on "23 May 1863" has been recorded. It was described as a "fixed barn engine" and in fact was first installed at Chalkpits Farm, the Englefield Estate home farm, being only moved to the sawmill in 1900. The two engines at Englefield are probably unique, and certainly historically important, survivals – these two makers would doubtless have built such engines in large numbers.

BUCKINGHAMSHIRE

5. Aylesbury, The Bifurcated & Tubular Rivet Co. Ltd., Mandeville Road (SP824125)

Photo taken/Reference: 30.7.1988/591-12
Type: Inverted vertical compound (enclosed)
Maker and Date: E. Reader & Sons Ltd., Nottingham. 1910
Cylinder dimensions: c.9ins/c.16ins x 6ins. Piston valves
Hp: c.140 **Rpm:** 450 **Psi:** 200
Flywheel diameter: 3ft 4ins
Service: Direct-coupled to DC generator (No. 2 Set)

Two apparently identical generator sets were installed here, new, in the older No.1 Factory, where they provided power for all the lineshafting, some electric lighting and a lift. They only stopped working, possibly around 1978, as the result of the failure of the superheater coils. At the time of the visit, the No.1 Set had already gone – removed to the Nottingham Industrial Museum, Wollaton Hall, in 1987; and this one also went for preservation, in 1989. The dynamos were by Siemens Brothers Dynamo Works Ltd., London & Stafford and rated at 100 kW. Two Lancashire boilers by Edwin Danks & Co. (Oldbury) Ltd., date 1946 and 1948, were out of use, and there was a more modern package boiler – in use, but details not recorded.

6. Loudwater, T.B. Ford Ltd., blotting paper works, Snakeley Mill (SU902903)

Photo taken/Reference: 27.10.1972/4436
Type: Uniflow
Maker and Date: Cole, Marchent & Morley Ltd., Bradford. No. 4308. 1921
Cylinder dimensions: 26ins x 27ins. Drop valves
Hp: 550 ihp **Rpm:** 150 **Psi:** 160
Flywheel diameter: 10ft 6ins
Service: Direct drive to mill and rope drive from flywheel (14 grooves in total) by 6 x $1^1/_4$ ins ropes to 150 kW generator

Referred to in *Stationary Steam Engines of Great Britain*, Volume 6 (see No. 42), this originated as a Cole, Marchent & Morley class C.E15 engine, with piston drop valves operated by a positive cam gear. However, this was apparently not very satisfactory and when in 1929 the piston and cylinder broke up, Galloways Ltd. of Manchester fitted a new cylinder with the Henry Pilling-Galloways oil hydraulic trip valve gear. A full description of this gear, in which the valves are operated by layshaft eccentrics and tripped by the governor is given in "Valve Gear 4. Pilling's Oil-hydraulic Gear", *SERG Bulletin*, Vol. 4, No.3, 1982. Superseded by the Hick, Hargreaves turbine installed in 1939, it appears to have remained on standby until 1971, when the works closed, but was scrapped around 1977.

4

5

6

7. Piddington, B. North & Sons (West Wycombe) Ltd., furniture manufacturers (SU810942)

Photo taken/Reference: 7.10.1974/19-10
Type: Horizontal side-by-side compound
Maker and Date: Marshall Sons & Co. Ltd., Gainsborough. No. 38817. 1903
Cylinder dimensions: 11ins/17½ins x 18ins. Slide valves; HP cylinder with automatic variable expansion slide valve
Hp: 100 **Rpm:** 108 **Psi:** 120
Flywheel diameter: 7ft 2ins
Service: Chain drives (two steps up) to alternator

This was an example of Marshall's compact side-by-side compound type, which could be supplied either with a boiler – often as a semi-portable undertype, or as here an "independent fixed engine". It is mounted on a steel frame and has both cylinders close together in a single casting and outside valves. In addition to the main slide valve, the HP cylinder has a cut-off valve controlled by a Hartnell governor and expansion gear. There was also an eccentric for a diagonal boiler feed pump. The engine appears to have given reliable service for many years, although reduced to standby duties when the factory was connected to the grid – probably some time before 1968. By then its sole purpose was to drive by Renold chains a 75 kW Crompton Parkinson Ltd. alternator. There appears previously to have been provision for a belt drive from the second motion shaft to a lineshaft above. The Lancashire boiler, also by Marshall, was fed by hand with wood waste. By 1980, the company's furniture manufacture had more or less ended, and the works was being used by a number of smaller companies. The engine remained as a rare and attractive survival which could still occasionally be seen running until 1985, and its removal in 1987 (although for private preservation) represented a sad loss.

CAMBRIDGESHIRE

8. Cambridge, Cambridge University Engineering Department (TL450576)

Photo taken/Reference: 12.5.1981/271-18
Type: Uniflow
Maker and Date: Robey & Co. Ltd., Lincoln. No. 40622. 1922
Cylinder dimensions: 12ins x 14ins. Drop inlet valves
Hp: 63.7 ihp **Rpm:** 180 **Psi:** 162 (Steaming figures from 1938 trial)
Flywheel diameter: 6ft 6ins
Service: Instructional

Installed by the University in the Engineering Department's thermodynamics laboratory for instructional and experimental purposes in 1922, this engine was fitted with various items of testing equipment including a rope brake dynamometer for measuring brake horsepower (7 ropes on the smooth flywheel rim). Typical of uniflow engines in being solidly built with enclosed counterbalanced double web crank and trunk guide as part of the main frame, the cylinder had double-beat drop valves operated by eccentrics on a skew gear-driven side shaft which also carried the governor. The valve gear was positive (without trip gear), the cut-off being controlled by the governor varying the position of the eccentrics in relation to the layshaft. A smaller, third eccentric operated a mechanism for operating decompression valves to prevent excess compression at times of insufficient vacuum. The surface condenser was below with a vertical air pump driven by a crank on the end of the crankshaft outboard of the flywheel. The engine had not been used for educational purposes since the late 1960s, but was still steamed on open days until the last time probably in July 1980. It was removed to the Strumpshaw Hall Steam Museum, near Norwich in 1982.

9. Histon, Cadbury Schweppes Foods Ltd., jam & preserve manufacturers (TL441626)

Photo taken/Reference: 19.5.1980/219-10
Type: Horizontal twin cylinder rotative pumping engine
Maker and Date: Sir William Arrol & Co. Ltd., Glasgow. 1902
Cylinder dimensions: c.12ins x 12ins. Slide valves
Rpm: c.11 (variable) **Psi:** c.80–90 **Flywheel diameter:** 4ft 8ins
Service: Pumping to hydraulic accumulator providing power for factory's lifts

The Victoria Works, established by Steven Chivers & Sons around 1874, was renamed The Orchard Factory in 1910, and purchased by Schweppes Ltd. in 1959. From 1896, the engineer had been Charles Lack, who had his own small engineering works in the nearby village of Cottenham, and it was he who was responsible for making and/or installing much of the machinery to be seen in the engine room – immaculately kept and containing an amazing assortment of engines and pumps. Remarkably, the small Arrol pumping engine was still in continuous use, providing for the relatively modest hydraulic power requirements of the factory's lifts. The small accumulator, which stood in the corner of the room, had a 10in ram and was loaded with 20 tons of sand. The pressure had been reduced from about 750 psi to 650 psi, probably some time in the 1970s. The pumps, in tandem behind the steam cylinders, had twin plungers linked by rods above and below. The engine may originally have had a throttle governor, but within living memory control had been solely by a connection to the accumulator. It remained in use until around June 1985, when all was being demolished around it, and three months later it was removed to the Bressingham Steam Museum in Norfolk.

7

8

9

10. Histon, Cadbury Schweppes Foods Ltd., jam & preserve manufacturers (TL441626)

Photo taken/Reference: 19.5.1980/218-29
Type: Inverted vertical single cylinder rotative fire pump
Maker and Date: Shand, Mason & Co., London
Cylinder dimensions: Unknown. Slide valve
Psi: 100
Service: Bucket pump drawing water from mains, forcing it at 80 psi into fire hydrant mains

Standing next to the hydraulic accumulator was another rare survival, this Shand, Mason fire pump, out of use before 1973, but remaining in situ until its removal for private preservation in 1985. Among other machinery, and still on standby, was a Belliss & Morcom compound, No. 3516, ordered by Chivers in 1908. This drove by vee belts an alternator, an AC induction motor, altered and installed to replace the original DC generator. Of the six Lancashire boilers, dating from the period 1897 to 1920, four were in use – the oldest by Thomas Beeley of Hyde, built in 1895, was installed second-hand in 1897, and had three flues, of which the bottom one had been blanked off.

CHESHIRE

11. Carrington, Shell Chemicals U.K. Ltd., Carrington Works (SJ735923)

Photo taken/Reference: 8.9.1980/234-5
Type: Horizontal single cylinder compressor
Maker and Date: Walker Brothers Ltd., Wigan. One of Nos. 48138 & 9. 1947
Cylinder dimensions: 25½ins x 30ins. Drop valves
Hp: 500 ihp **Rpm:** 110 **Psi:** 160
Flywheel diameter: 12ft
Service: Compressing process gas and/or ethylene cycle gas in Ethylene Plant No.1

Walker Brothers had been important builders of compressors since around 1870 and the six here were possibly the last they built before going out of business. They were ordered by Petrocarbons Ltd., Manchester – four of them (Walker Nos. 48316–48319) on 26 July 1946. (The other two are as yet untraced in Walkers' records.) Standing side by side in the same long building, these massively built engines were an impressive sight, although effectively derelict, having been out of use probably since 1969. At the steam end they were identical, with enclosed single web cranks, cylinders of the same size, and drop valves driven by eccentrics from the bevel gear driven layshaft. A Whitehead governor controlled the inlet valve trip gear. The exhaust steam at c.20 psi went to process. The gas cylinders, connected to the rear of the steam cylinders by heavy struts, differed in size. Four were 21½ins diameter with a capacity of 1900 cfm and raising pressure from 2 to 10 atmospheres absolute. The others (on Nos. 48138-9 and including that on the engine illustrated) were 39ins diameter with a capacity of 3700 cfm, a suction pressure of 1.05 atmospheres and discharge of 4.1 atmospheres.

12. Congleton, John Pattison (Congleton) Ltd., textile printers, Dane-in-Shaw Mill (SJ883620)

Photo taken/Reference: 17.4.1975/28-27
Type: Diagonal twin cylinder
Maker and Date: Mather & Platt, Manchester. 1891
Cylinder dimensions: c.6ins x c.6ins. Slide valves
Flywheel diameter: 2ft 6ins
Service: Driving drying cylinders – by five vee belts from small pulley

This was typical of the type used for driving textile finishing machinery, with a single double web crank for the opposing inclined cylinders and a single eccentric for both slide valves. Slightly more unusual, however, were the forked connecting rods to allow room for the bush-guided forward extensions of the piston rods. The drive was taken from the second motion shaft by a complicated system of chains and belts to a bank of drying cylinders, which presumably took the engine's exhaust steam. It had been out of use, probably since around 1968, and its previous history was unknown. It had doubtless given good service since its assumed date of manufacture (the figure 1891 was given with the maker's name on the valve covers). The only means of speed control had been by hand-adjustment of the steam inlet – apparently required quite often, given the need to keep the cylinders turning at a constant speed and variations in the steam demand of other parts of the works and boiler pressure changes. Another (presumably similar) engine had apparently already been removed for preservation; this one had gone by 1978 and all manufacturing at the mill ceased in the early 1980s.

13. Hazel Grove, A. Barton & Sons Ltd., wadding manufacturers, Albion Mills (SJ918874)

Photo taken/Reference: 24.4.1978/104-16
Type: Horizontal cross compound
Maker and Date: S.S. Stott & Co., Haslingden. 1903
Cylinder dimensions: c.10ins/c.20ins x 3ft. Corliss valves (HP); slide valve (LP)
Hp: 200-250 **Psi:** 160
Flywheel diameter: 10ft
Service: Belt drive

It has so far been impossible to discover a great deal about the history of this engine or its working conditions. However, it was said to have been installed here second-hand from the Chorley Rail Wagon Company in 1920 .When seen in 1978, it had been out of use for more than ten years, but it was taken to the Anson Steam Museum, Poynton in 1984, soon after the works closed. No details of the machinery driven are to hand, but it was a solidly built engine, with drive by a canvas belt from a pulley 6ft 6ins diameter and 24½ins wide, adjacent to the flywheel. The belt, 17¾ ins width, can be seen coiled up on the engine room floor. The governor, similar to a Watt type, but with central weight, controlled the HP trip gear, and the vertical air pump beneath the engine room floor was driven by a linkage and bell crank from the LP tailrod slide.

12

14. Macclesfield, The Barracks Fabrics Printing Co. Ltd., Lower Heys Mill (SJ919743)

Photo taken/Reference: 24.6.1993/723-28
Type: Diagonal twin cylinder
Maker and Date: Robey & Co. Ltd., Lincoln. No. 32243. 1913
Cylinder dimensions: 9ins x 14ins. Slide valves
Rpm: Variable **Psi:** c.40
Flywheel diameter: 3ft 6ins
Service: Driving calender

Few engines of this type survived beyond the 1970s, and this was probably the last one to remain in use, only stopping in 1995. Using steam at 40 psi, reduced from the boiler's 100 psi, this seemed a typical example, with 4-bar crosshead guides, and geared down (c.2.25:1) to the central steel roll of the calender by Bentley & Jackson Ltd., Bury. The speed of working was given as 20 metres of cloth per minute. Another adjacent calender (not steam engine driven) was of similar date (or at least to that of the Robey engine), having been built by Jackson & Brothers in 1911. Probably they were installed here second-hand – the Robey engine appears to have been first used in one of the large number of works operated under the aegis of the Calico Printers' Association. The mill had been long established as a cotton manufacturing mill, but weaving (in a very large shed added around 1867) apparently ceased around the early 1930s, by which time the Barracks company, operating purely as textile finishers, was in occupation. In 1996, the Company re-located its production to Lancaster and the engine went for preservation at the Dingles Steam Village, near Lifton in Devon.

15. Marston, Ingram Thompson & Sons Ltd., Lion Salt Works (SJ671753)

Photo taken/Reference: 24.4.1978/105-5
Type: Horizontal single cylinder
Maker and Date: c.1900? Rebuilt Marcus Allen, Manchester. 1910
Cylinder dimensions: 8ins x 16ins. Slide valve
Flywheel diameter: 5ft
Service: Brine pumping from depth of 150ft into reservoir, then gravity feed into evaporating pans

An important survival, the Lion Salt Works was the last in the U.K. to produce salt by the open pan boiling of pumped brine. Operations ceased in 1986, but the site is now being preserved. By the 19th century, Cheshire brine was being raised by steam engine driven pumps – initially by beam engines, but finally perhaps more usually by horizontal engines, the arrangements at the Lion Works probably being typical. The engine, in a wooden shed, is a very basic slide valve engine, possibly dating from around 1900, but installed here as late as 1937. A (c.5½:1) reduction geared second motion shaft had a disc crank, from which a wooden pitman beam (stroke 42ins), passing through the end wall, operated the pump bob supported on two closely spaced parallel brick walls. By the 1960s, an electric motor had been connected to the engine crankshaft, which with the pump gear was used as a standby to the main pump. Across the yard, in its own house, is a small Cornish boiler by William Lord of Bury, dated 1891, but apparently also not put in here until 1937.

14

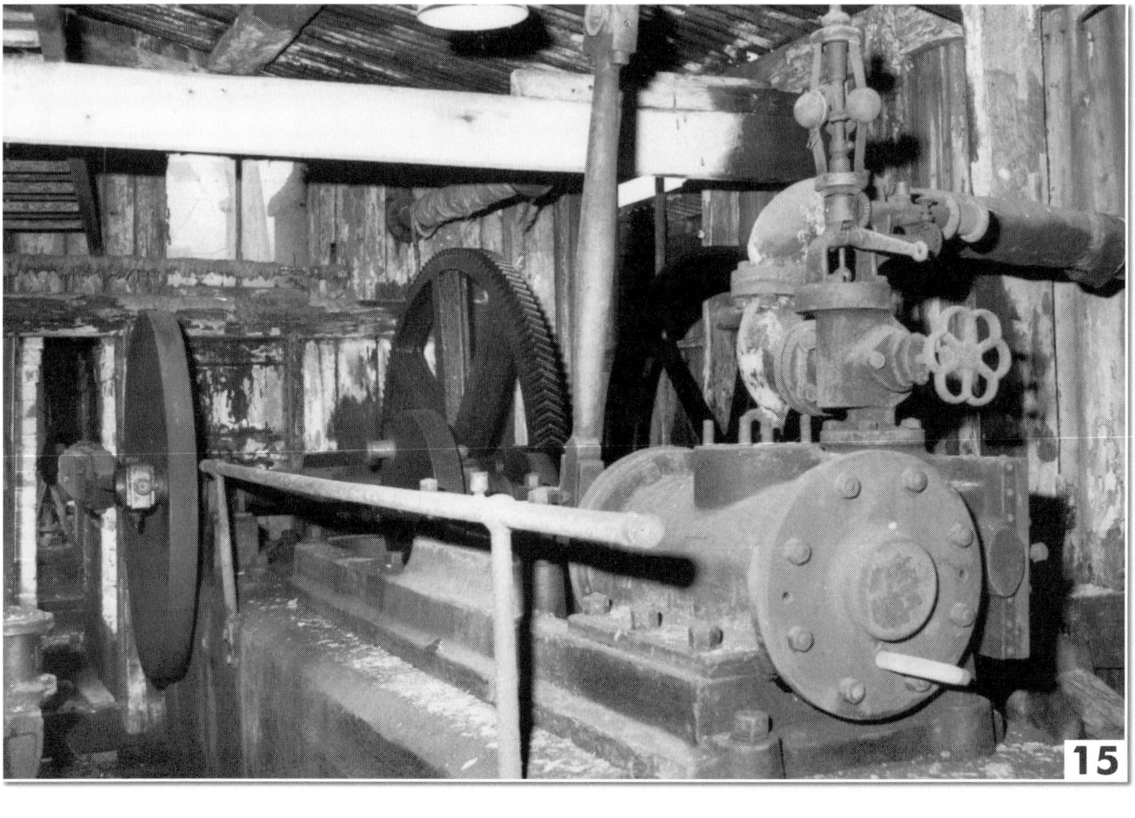

15

16. Northwich, British Waterways Board, Northwich Yard (SJ657735)

Photo taken/Reference: 24.4.1978/103-37
Type: Horizontal single cylinder
Maker and Date: Edwin Foden Sons & Co., Sandbach. 1870s?
Cylinder dimensions: 12ins x 20ins. Slide valve; Stephenson reversing link
Hp: 20? **Psi:** c.80?
Flywheel diameter: 6ft 1½ins
Service: Driving four slipway haulage winches

This interesting installation survived until 1987, when finally, sadly, the engine was removed to the British Waterways Museum, Gloucester Docks. The yard, on the River Weaver, was established in 1876 by Samuel Bracegirdle, who may possibly have installed the engine and three at least of the winches. The fourth was added later to accommodate longer boats. It was owned by the Liverpool Lighterage Company from c.1900 to 1972, when it was taken over by British Waterways, but the use of the slipway had possibly been discontinued by then. All was still more or less intact in 1978, although the slips were picturesquely overgrown with grass and there was no sign of the wheeled carriages. The engine looks right for the supposed date, being solidly built with bent crankshaft and 4-bar crosshead guide. A slide valve with Stephenson link reversing, together with a Porter governor controlling the steam inlet, were fitted. The engine drove by reduction gearing (approximately 5:1) a long shaft from which there was further reduction to each of the winches along its length. The engine probably stopped work in the 1940s, when the boiler (no longer present) became inoperable. However, it was superseded by a DC electric motor built by British Thomson Houston Co. Ltd., Rugby, which was connected to the winch shaft via two stages of reduction gearing. With its controller, this came second-hand from a Liverpool Corporation electric tramcar.

17. Northwich, ICI Ltd., Mond Division, Wallerscote Works, Winnington (SJ641745)

Photo taken/Reference: 2.5.1985/481-36
Type: Horizontal single cylinder compressor
Maker and Date: Walker Brothers Ltd., Wigan. No. 24545. 1920
Cylinder dimensions: 21ins x 3ft. Drop valves
Psi: 145?
Flywheel diameter: 14ft
Service: Part of soda production plant

The following four horizontal engines, still in situ in 1985, were ordered from Walker Brothers by Brunner Mond & Co. Ltd. for their Solvay process soda production plant, which started operations in 1926:
Nos. 24545 & 24546 1920 Compressor cylinder 32ins dia
No. ? 1920 Compressor cylinder 32ins dia. This engine was rebuilt as No. 30950 in 1929 with a 50ins dia. vacuum pump cylinder
No. 24973 1921 Vacuum pump cylinder 50ins dia.
Of massive build, they were similar in design and appearance, each with compressor or vacuum cylinder (with tailrod) in tandem behind the steam cylinder. They had 21ins x 3ft stroke cylinders with drop valves operated by eccentric rods from a bevel gear driven side shaft. A Whitehead governor controlled the steam valves, using Doerfel's (positive) valve gear. The compressor cylinder valves had their own eccentric rod from a short shaft at 90° to, and driven by bevel gearing from, the main eccentric shaft. (The valves of the vacuum pumps were self-actuating.) They were non-condensing engines, exhausting to LP turbines or process. The compressors probably stopped work c.1977/8, the vacuum pumps c.1982 and (No. 24973) 2nd October 1984, when the plant was finally shut down. No. 24973 was removed to Trencherfield Mill, Wigan for preservation in 1985, but the others were scrapped.

16

17

18. Stalybridge, John Hargreaves (Collyhurst & Stalybridge) Ltd., Tameside Paper Mills (SJ967989)

Photo taken/Reference: 6.8.1988/603-37
Type: Inverted vertical compound (enclosed); Inverted vertical twin cylinder (enclosed)
Maker and Date: Ashworth & Parker Ltd., Bury. No.197 - 1910; No.863 - 1921/2
Cylinder dimensions: 12ins/18ins x 7ins; 12ins x 6ins. Piston valves
Hp: c.55 bhp; 80 bhp **Rpm:** 30–240; 50–400 **Psi:** 140
Flywheel diameter: 5ft; 4ft 6ins
Service: Rope drive to wet end of paper machine, plus six drying cylinders

This was one of the last paper mills to be driven by a steam engine (only one other – Frogmore Mills, in Hertfordshire – is known to have outlasted it – see No.74). It was thought that both these engines had been installed second-hand, probably in the late 1930s or 1940s, the compound being standby only. The twin cylinder engine, on the right, seen at work, had run more or less continuously 24 hours a day, 6 days a week for the previous 8 years, and for a long period before that was said to have worked 7 days a week. As was typical with paper mill drives, speed was variably controlled by a Pickering governor driven via a moveable belt between two cone pulleys. The exhaust steam was used to heat the drying cylinders. The two coal-fired Lancashire boilers were by Daniel Adamson & Co., of Hyde, using mechanical stokers. Sadly, by the year 2000, the engine had been relegated to standby duties only.

19/20. Wincham, Ingram Thompson & Sons Ltd., Sunbeam Salt Works (SJ679754)

Photo taken/Reference: 24.4.1974/12-37 & 12-30
Type: Horizontal single cylinder
Maker and Date: Abraham Lord, Rochdale. 1885 (reputedly)
Cylinder dimensions: c.10ins x 20ins. Slide valve
Hp: c.9 ihp **Rpm:** 35 **Psi:** 30
Flywheel diameter: 5ft 8ins
Service: Brine pumping

(Figures for working conditions derived from Arthur Bebbington's report reproduced in the *AIA Bulletin* 3.3, Oct 1976.) A small outlying brine pumping station, operated by the same company as the Lion Salt Works, this was completely derelict when seen in 1974, and is believed to have stopped working in 1967. A few months later, the engine was stolen by scrap metal thieves. It was very similar to that at the Lion works, the main difference being the Porter governor driven by return crank and bevel gearing (compared with the belt driven Pickering governor of the Lion works engine). The pump drive was also similar, with herring-bone reduction gearing (5:1 in this case) to the second motion shaft and a pitman beam to the pump bob – here supported at ground level. The brine, pumped from a depth of about 125ft, passed via a receiver into a reservoir from which it flowed by gravity to the pans half a mile away. The photograph of the exterior gives some idea of its ramshackle appearance, but it may not have looked very different when in use!

CORNWALL

21. Carthew, Greensplat Clayworks (F.W. Berk & Co. Ltd.) (SW997554)

Photo taken/Reference: 27.9.1962/1860
Type: Cornish beam pumping engine
Maker and Date: Possibly West & Sons, St. Blazey Foundry. c.1850
Cylinder dimensions: 30ins x 9ft. Cornish double-beat valves
Spm: 9
Service: Pumping clay slurry

Quite small for a Cornish pumping engine, this had been built as a rotative engine and was first erected at Bunney Mine, a mile or two to the north, before being purchased (apparently for £65) by Greensplat China Clays in 1894. It was erected at Greensplat in 1897, without its flywheel. According to the 1933 Cornish engines survey, it had an unequal beam, the stroke being 8ft in the shaft, and a single cataract. It was also apparently fitted with an early form of "St. Austell governor", in which the cataract cock was linked to a float in the shaft sump. With 240ft of rods in the shaft, it raised 500 gallons of clay slurry per minute at 9 spm. It was the last Cornish engine to work in Cornwall, only stopping on 23 February 1959. When visited in 1962, the clay works had recently been taken over by the chemical manufacturers, F.W. Berk. After remaining in situ for another ten years, the engine was removed to Wendron Forge for preservation at a reported cost of nearly £8000. But preserved in the open, and with its pump rod waving around above ground, this was never a very satisfying exhibit.

19

20

21

22. Constantine, Tresahor Granite Quarry (SW742312)

Photo taken/Reference: 23.9.1987/564-5
Type: Mast crane winder (diagonal twin cylinders)
Maker and Date: Unknown
Cylinder dimensions: c.6ins x 1ft. Slide valves; Stephenson link reversing
Service: Winding mechanism for the crane. Drum in two halves, each 29ins in width, diameter 24ins, driven by reduction gearing (possibly $4^1/_2$:1).

The crane is believed to have been working by compressed air from c.1900 to the late 1930s, the quarry being operated for some of this time by John Freeman & Co. Ltd. The compressed air was piped c.1000 yards from an electrically-driven compressor. All that was left in 1987 was the winding gear, or part of it at least. The crane itself is thought to have fallen into the flooded quarry. Of the two sides of the "engine", all was missing on one side, as was the crankshaft and the pinion gear it carried. Remaining was most of the framework, the drum, and parts of the engine on the other side, including cylinder, piston rod, crosshead, guide bars and connecting rod.

23. Luxulyan, Treskilling China Clay Works (SX035577)

Photo taken/Reference: 29.5.1978/113-12
Type: Horizontal twin cylinder haulage
Maker and Date: Unknown
Cylinder dimensions: c.9ins x 18ins. Slide valves; Stephenson link reverse
Service: Haulage of waste sand to tip. The two drums were each 50ins diameter and $16^3/_4$ins wide

The two sides of this engine had centre lines about 8ft apart with the two drums between. These were driven from a pinion on the crankshaft to a gear wheel on the end of one of the drums (ratio 3:1). The adjacent vertical boiler carried the stencilled wording: PETER DIXON & SON Ltd. TRESKILLING CHINA CLAY WORKS, this being the paper-making company which had operated the works since taking it over in 1924. However, it is perhaps more likely that the engine had been installed by their predecessors, the Treskilling China Clay Co. Ltd. Located directly in line with the top of the overgrown waste heap it had helped to create, this interesting machine had become derelict by the mid 1960s and by 1983 it had been removed to Morwellham Quay for preservation.

22

23

24. Penzance, Penzance & District Steam Laundry Ltd. (SW474303)

Photo taken/Reference: 31.3.1980/206-3
Type: Horizontal single cylinder
Maker and Date: J.J. Lane Ltd., London. Possibly 1901
Cylinder dimensions: c.12ins x 15ins. Slide valve
Psi: 100
Flywheel diameter: 5ft
Service: Belt drive from 42ins diameter pulley to lineshaft, thence to laundry machinery

When this engine was inspected in 1980, the laundry had recently closed, but unfortunately all the machinery (doubtless including the usual washing machines and hydro-extractors etc.) had been sold by auction and removed a few days earlier. The owner (Mrs Boase) understood that the engine had been installed when the laundry opened in 1901, and it seemed quite possible that it was then new. J.J. Lane was a company well-known for its laundry machinery and clearly quite capable of providing the solidly built and reliable engines necessary to drive it. This example had given more than fifty years' service and was not removed until the 1990s, when the buildings were demolished. It was later seen at the Shire Horse & Carriage Museum, Treskillard, near Redruth.

25. Pool, East Pool & Agar Mine – Michell's Shaft (SW674415)

Photo taken/Reference: 27.9.1960/775
Type: Cornish rotative beam winding engine
Maker and Date: Holman Bros., Camborne. 1887
Cylinder dimensions: 30ins x 9ft. Cornish double-beat valves; Gooch reversing link
Rpm: 17 (27-30 des.) **Psi:** c.40
Flywheel diameter: 22ft
Service: Wound men and ore from 1,490ft. Two drums 9ft diameter. Rope speed up to 1,000 ft/min

The north shaft winder (the "North Whim"), designed by F.W. Michell and erected by the East Pool company, was the only rotative beam winder built by Holman Brothers and possibly the last such engine to be built anywhere. The valves are driven by two eccentric rods through a Gooch reversing link and rocking shaft. A separate expansion valve (giving perhaps c.30-40 per cent cut-off), driven by a shaft via a cam and roller, can be operated by a foot pedal. The two drums were placed directly on the crankshaft with the flywheel between them. The engine stopped work in 1921 as a result of an underground collapse in the nearby Engine Shaft, but some years later the Cornish Engines Preservation Society helped to secure its survival, after which it was taken over by the National Trust. (Since this photograph was taken, a replica boiler house has been added.)

24

25

26. Pool, East Pool & Agar Mine – Taylor's Shaft (SW675417)

Photo taken/Reference: 24.9.1987/564-35
Type: Cornish beam pumping engine
Maker and Date: Harvey & Co. Hayle. 1892
Cylinder dimensions: 90ins x 10ft. Cornish double-beat valves
Hp: 425 **Spm:** 5 **Psi:** 50
Service: Pumping by seven lifts from a depth of 1,700ft. Plunger pole pumps 16ins (at lowest level) and 18ins x 9ft stroke. Capacity 450 galls/min

Following the underground collapse at East Pool in 1921, the company (by then East Pool & Agar Ltd. as a result of an amalgamation in 1897) decided to continue operations by sinking Taylor's Shaft nearby. (M.T. Taylor was the mine superintendent). The 1892 engine, built by Harveys and designed by Nicholas Trestrail, was erected here in 1924 – the last installation of a beam pumping engine to take place in Cornwall. A second-hand engine, it had been built for the Highburrow East Shaft of the Carn Brea Mines, where it remained in use until c.1917. Heavily built, it gave very reliable service until 1954, although during its last nine years with the pumping depth reduced to 1,034ft, its sole purpose was to prevent the adjacent South Crofty working from being flooded. It was then presented to the Cornish Engines Preservation Society by the American engineering historian, Mr. Greville Bathe, before being eventually taken over by the National Trust. (The photograph shows the cylinder and top nozzle: governor, steam and equilibrium valves.)

27. Pool, South Crofty Ltd., South Crofty Mine – Palmer's Shaft (SW669413)

Photo taken/Reference: 14.4.1974/11-22
Type: Horizontal twin cylinder winder
Maker and Date: E. Wilkes, Pelsall Foundry, near Walsall. Probably c.1907
Cylinder dimensions: 18ins x 3ft. Slide valves; Stephenson? link reversing
Psi: 100
Service: Winding from shaft. Parallel drum, 7ft diameter

This engine was installed as a replacement of a very old single cylinder winder, and was probably new when installed c.1907 (its builder had been known as J. Wilkes before 1901). Also put in was a single Lancashire boiler, 28ft x 7ft 6ins, working pressure 100 psi. Winding from Palmer's Shaft was stopped in 1921, but the engine remained, increasingly derelict, until its eventual removal or scrapping around the mid 1970s. Unfortunately, the wooden engine house, already semi-ruinous, was set on fire c.1970/1, and when photographed the engine was half buried in debris. For its date, it was perhaps of somewhat antiquated design, an unusual and not very convenient feature being the entry of the steam pipes through the flat valve chest covers. The drum, 7ft diameter by 5ft wide, which had a maker's plate: RATCLIFFE & SONS, HAWARDEN IRON WORKS, was almost certainly a narrower replacement of the original, probably to give better coiling on the drum. Palmer's Shaft was reconstructed 1949/1950, but instead of using the Wilkes engine, the smaller and slower geared winder in the nearby Bickford Shaft engine house was refurbished, with the rope being turned about 90° by an angle sheave. Some of the pieces of this engine, built by John Wild & Co. Ltd., Oldham in 1895, were still lying around in its house.

28. Pool, South Crofty Ltd., South Crofty Mine – Robinson's Shaft (SW668413)

Photo taken/Reference: 1.4.1980/207-33
Type: Horizontal twin cylinder capstan
Maker and Date: Harvey & Co., Hayle. 1882
Cylinder dimensions: 12ins x 18ins. Slide valves; Stephenson reversing links
Flywheel diameter: 7ft
Service: Shaft and pitwork maintenance etc. Drum c.4ft 6ins diameter

This worm capstan was apparently a standard design by Loam & Son, who were closely linked with the Harvey company at this time. It had cross-coupled twin cylinders, the direct-coupled wormshaft driving to a wheel at one end of the drum. It was presumably built for Tregurtha Downs Mine, Marazion, and from there transferred with the Cornish pumping engine to South Crofty in 1903. (The latter had been erected third-hand at Tregurtha Downs by the Loams in 1882–3.) It may have remained workable by compressed air until the 1980s before South Crofty's final closure, and with the pumping engine (see *Stationary Steam Engines in Great Britain*, Volume 7, No. 1) remains in situ.

Another very similar engine was probably built in 1881 for Tresavean Mine before being used at Wheal Grenville (Fortescue's Shaft) from 1892, and then in 1922 finally installed at New Cook's Kitchen Shaft, South Crofty, from where it was removed for preservation in 1985.

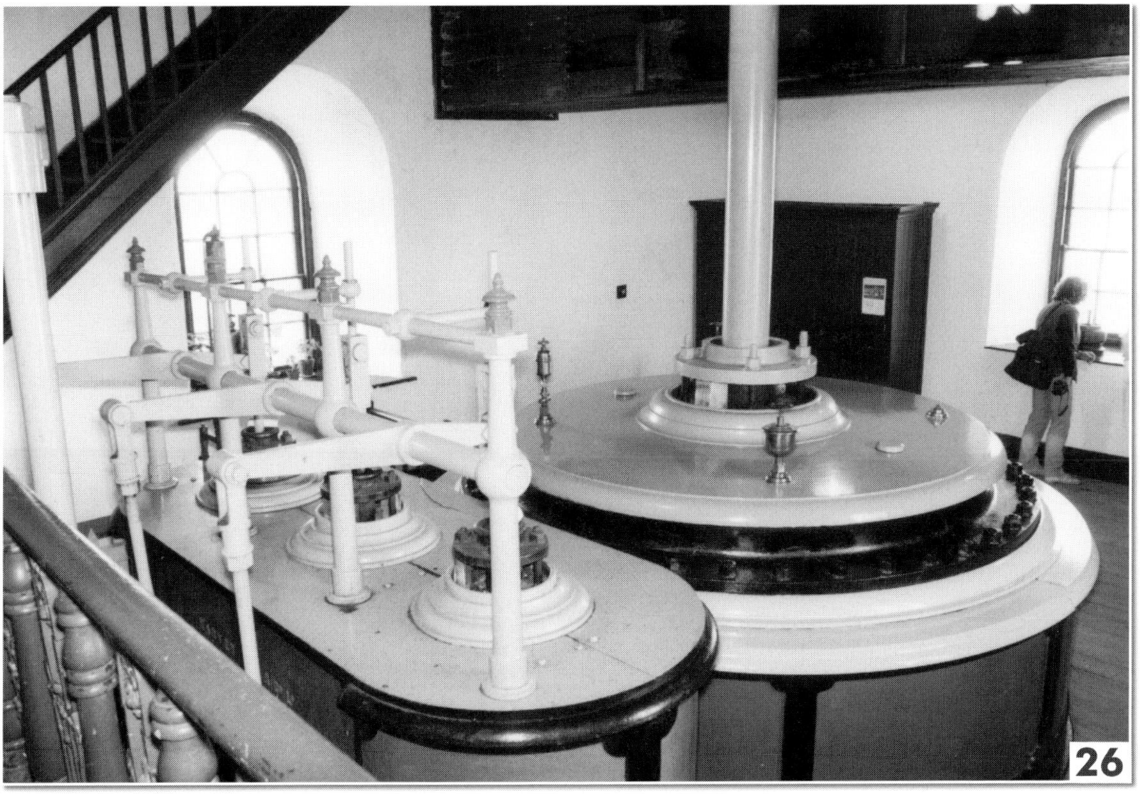

26

27

28

29. Redruth, Devenish Redruth Brewery Co. Ltd., Foundry Row (SW695422)

Photo taken/Reference: 2.4.1980/208-35
Type: Horizontal single cylinder
Maker and Date: G.Adlam & Sons, Bristol. c.1870s
Cylinder dimensions: c.6ins x 10ins. Slide valve
Psi: 80?
Flywheel diameter: 46½ins
Service: Driving hoist by belts from divided pulley 16ins diameter to shafts above

There were two horizontal engines here – possibly installed as early as the 1870s, presumably by the Redruth Brewery Company. (Later the brewery was acquired by the Weymouth brewers, J.A. Devenish & Co., and the company became the Devenish Redruth Brewery Co. Ltd. in 1960.) The smaller of the two was the hoist engine, located at the top of the brewery, and built by the well-known brewery engineers, George Adlam. This tiny engine, on a substantial concrete base, had been out of normal use possibly since as long ago as the 1950s. The other engine, on the ground floor, thought likely to be by the Redruth Foundry Co. Ltd., remained in use until around 1984, although for its last ten years its sole function had been to drive a 3-throw liquor pump (also by Adlam). There were two Lancashire boilers by Edwin Danks & Co. (Oldbury) Ltd., 1928 and 1939, converted to oil-firing in 1965. The larger engine was removed in 1987 and is now privately preserved, but it was only after the brewery (by then under different ownership) was closed in 2004 that the Adlam engine also was removed – apparently to be preserved by BTB Brau-Technologie GmbH, in Germany.

29

30. St. Austell, English Clays Lovering Pochin & Co. Ltd., Trethowel Clayworks (SX010539)

Photo taken/Reference: 5.4.1993/712-19
Type: Two horizontal single cylinder engines (coupled)
Maker and Date: Installed c.1920s
Cylinder dimensions: c.8½ins x 12ins (one cylinder). Slide valve
Flywheel diameter: 5ft 2ins
Service: Coupled to a winding drum c.2ft diameter and c.2ft wide used for winding clayworks sand to waste heaps

The two unmatched engines, presumably installed by the Gears Clay Company, then operating the works, were coupled together and geared down (c.3:1) to a second motion shaft connected to the winding drum through a dog clutch. One of the two cylinders appeared to be missing (although reportedly somewhere in the vegetation), but most of the basic ironwork of the two engines was still present, although some, such as the second motion shaft with spur gear and coupling shown in the right foreground, lying detached. Close by were the remains of another rather larger horizontal engine which drove a pump by flat rods, and the housing of a Holman Bros. Cornish boiler of 1900, which had been removed to Levant Mine in 1990. All is believed to have stopped work c.1935.

31. St. Dennis, English Clays Lovering Pochin & Co. Ltd., Parkandillick Clayworks (SW947568)

Photo taken/Reference: 21.9.1987/562-21
Type: Horizontal twin cylinder haulage
Maker and Date: Davey, Paxman & Co. Ltd., Colchester. c.1890s?
Cylinder dimensions: 12ins x 14ins. Slide valves
Flywheel diameter: 5ft
Service: Skip winding engine. For hauling china clay waste to tip. Two drums, 4ft diameter and 18ins wide, clutchable independently, driven via single (3:1) reduction gearing

A second-hand Cornish pumping engine by Sandys, Vivian & Co., Copperhouse Foundry was installed here by the St. Dennis and Parkandillick Clay Company Ltd. in 1912 (see George Watkins, *The Steam Engine in Industry - 2. Mining and the Metal Trades*, 1979). The haulage engine, also presumably second-hand, was probably put in at the same time, with an added or replacement drum section, completely independent and plated: BEVER DORLING & Co. Ltd., BOWLING IRON WORKS, BRADFORD. With cylinders in a single casting each side of the valve chests, and a bent crankshaft, this was apparently typical of the small haulage engines being built by Davey, Paxman from the 1880s. In the same room are a locomotive type boiler and a small vertical boiler, presumably for times when the main H. & T. Danks Lancashire boiler was not in use. Together with the pumping engine, which stopped work in 1953, it is preserved and can be operated by compressed air.

32. St. Just, Geevor Tin Mines Ltd., Geevor Mine – Victory Shaft (SW375345)

Photo taken/Reference: 3.4.1980/211-21
Type: Horizontal twin cylinder winder
Maker and Date: Probably pre-1906. Installed (second-hand) 1923
Cylinder dimensions: 18ins x 3ft. Slide valves; Stephenson link reverse
Psi: 75
Service: Rope changing etc. in shaft from 1954, using compressed air. Two parallel drums, 6ft diameter

This engine was installed c.1906 at Wheal Vor (Borlase's Shaft), possibly second-hand, and Robey & Co. Ltd., Lincoln has been suggested, without much evidence it seems, as the possible builder. After operations at Wheal Vor ceased c.1910–11, the engine was sold to Geevor. The sinking of Victory Shaft started in 1919, but erection of the engine was only completed in 1923. Originally with a 9ft diameter double drum, the existing two drums, one being declutchable by a hand wheel, were fitted by Bartle & Sons, Carn Brea during the Second World War. Installation of an electric winder in 1954 brought its use for ore raising to an end. Following final closure in 1991, the mine became a museum, which at the time of writing appears to be facing an uncertain future.

30

31

32

33. St. Just, Levant Mine (SW368345)

Photo taken/Reference: 18.5.1982/GC1982.30.13
Type: Cornish beam winding engine
Maker and Date: Harvey & Co., Hayle. 1840. Rebuilt 1863.
Cylinder dimensions: 27 7/8ins x 4ft. Cornish double-beat valves
Rpm: c.28 **Psi:** c.25-30
Flywheel diameter: 12ft 2ins
Service: Winding ore from Skip Shaft (until 1930). Two drums, 4ft 10ins diameter. Maximum rope speed 400 ft/min

When Levant Mine closed in 1930, this historic engine was still in use, but five years later it was on the point of being scrapped. The need to prevent this and purchase the engine for preservation prompted the formation of the Committee, which in 1943 became the Cornish Engines Preservation Society. In 1967, this and the other engines by then belonging to the Society were handed over to the National Trust. In 1992 the engine was restored to steam, but unfortunately the perhaps excessive guarding apparently found necessary since this photograph was taken has detracted from the engine's appearance. The four valves, steam and exhaust at top and bottom of the cylinder, are operated by a single horizontal arbor connected to two eccentrics – one for each direction of the engine's rotation. The valves were originally operated by a plug rod, but all was changed after a rebuild by Hocking and Loam in 1863, which included a new cylinder and flywheel. At some point, also, the original geared flat rope drum was replaced by the existing drums, direct-driven.
(Photograph by George Cooper. Courtesy of English Heritage.)

33

34. St. Mewan, Burngullow China Clay Kilns (SW979525)

Photo taken/Reference: 29.5.1978/113-25
Type: Inverted vertical twin cylinder haulage
Maker and Date: Unknown
Cylinder dimensions: c.7½ins x probably 8ins. Slide valves; Stephenson reversing links
Service: Driving winding drum, 20ins dia x 26ins width, by reduction gearing (c.3⅛:1). Presumably for hauling china clay waste to nearby tips

This unit consisted of three independent parts – vertical boiler; engine; and drum section – bolted to an RSJ framework. It was mounted on wheels but had the appearance of never having moved. The engine had two separate cylinder and valve chest castings with a single supporting frame, and the bent crankshaft carried a clutchable pinion meshing with a 4ft diameter gear wheel on the drum shaft. When seen in 1978, disuse and exposure to the elements over perhaps many years appear to have left their mark, and the engine was suffering from rust and decay, and missing parts. It did not remain here much longer, being moved to Morwelham Quay, perhaps about a year later.

35. St. Stephen-in-Brannel, Goonvean & Rostowrack China Clay Co. Ltd., Goonvean Clayworks (SW949553)

Photo taken/Reference: 7.5.1990/648-5
Type: Cornish beam pumping engine
Maker and Date: Harvey & Co., Hayle. 1863
Cylinder dimensions: 50ins x 10ft. Cornish double-beat valves
Service: Pumping clay slurry. Two stages of plunger pumps in 365ft deep shaft. Unequal beam, pump stroke being 8ft 9ins. Also, formerly pumped from two other shafts by means of flat rods

The engine was built to a design by Thomas James for Penhalls Mine, moving via Trevaunance Mine to Gooninnis Mine before installation for the Goonvean China Clay Company in 1910. It had a double plug rod and two cataracts and was fitted with a "St. Austell governor" – a standard feature of clay pumping engines – to prevent fluctuations in the stroke. Steam was provided by a single Lancashire boiler. The beam is a 1928 replacement by Holman Bros., thought to be possibly the last ever to be cast for a beam engine anywhere in the world. Centrifugal pumps were installed in 1956, but the engine – becoming increasingly derelict – has remained here ever since, despite severe storm damage to the engine house in 1990 and various so far abortive plans for its removal.

34

35

36. St. Stephen-in-Brannel, sawmill(?), Treviscoe (SW936560)

Photo taken/Reference: 29.5.1978/113-33
Type: Horizontal single cylinder
Maker and Date: Unknown
Cylinder dimensions: 12ins x 15ins. Slide valve
Flywheel diameter: 5ft 6ins
Service: Gear drive to sawmill machinery?

This site represented something of an archaeological enigma, the first difficulty being that of actually finding it in the woodland depths in which it was hidden. The engine was said to have driven a sawmill, of which little but a few foundations could easily be seen. It was of the simplest slide valve type, and had a substantial tree growing through its middle with a covering of ivy, some of which was removed for photography. It had a single slipper guide and two eccentrics, but no valve rod(s) or sign of any reversing gear. A single pinion on the end of the shaft represented the first stage (all that remained) of the reduction gearing to a second motion shaft. The building is shown on the 1st edition 25-inch map dated 1879, but no other information on the history of the site has come to hand. The engine was taken to the Wheal Martyn museum site in 1983/5.

37. Torpoint, Ministry of Defence, Thanckes Oil Fuel Depot (SX432562)

Photo taken/Reference: 1.6.1978/116-27
Type: Horizontal duplex triple-expansion non-rotative pumping
Maker and Date: Worthington Pump Co. Ltd.*, London (built at Newark). Nos. 3003 & 3004. 1917
Cylinder dimensions: 12ins/19ins/30ins x 2ft. Semi-rotary slide valves; two auxiliary cut-off valves for each HP cylinder
Spm: Up to 35 **Psi:** 60
Service: Fuel oil pumping. Original duty to pump 270 tons/hour, discharge pressure 85 psi. Pump rams 12½ins dia x 2ft stroke

*This was the holding company owning the shares of James Simpson & Co. Ltd.

These were among the group of four engines (Nos.3001-4) originally ordered by the Admiralty for the Oil Fuel Depot at Port Edgar in Scotland. Nos. 3003-4 were moved here around 1920; and when Port Edgar closed in 1937, the others went to Lyness in the Orkney Islands, where they were erected with another (new) one – Worthington-Simpson No. 3105. By 1978, the engines were exhausting to atmosphere, their condensers having been removed, and the wooden cooling tower taken down (its maintenance by creosoting had been found too expensive). In 1980, the two large original Lancashire boilers (oil-fired) were scrapped, superseded by three second-hand vertical boilers by Cochran & Co., built 1951, but the pumping engines continued in use until March 1992, when stopped by a cracked cylinder. By then, they were used not for delivery but, because they could reverse the flow, for pulling the oil back along the line.

36

DERBYSHIRE

38. Bretby, Bretby Brick & Stoneware Co. Ltd., Newhall (SK281215)

Photo taken/Reference: 8.3.1982/322-3
Type: Horizontal single cylinder
Maker and Date: Walker Brothers (Wigan) Ltd., Pagefield Ironworks, Wigan. No.23974. c.1920?
Cylinder dimensions: 18ins x 30ins. Piston valve; Stephenson reversing link
Rpm: 70 **Psi:** 80?
Flywheel diameter: 12ft
Service: Driving brickmaking machinery and claypit haulage system until 1957/8; then stoneware machinery 1960–69 (by belt from 7ft diameter pulley to lineshaft)

In 1898, Bretby No.2 Pit was leased by the Earl of Caernarvon to Messrs Perkins, Warren & Barnes, who established Bretby Brick Works on this site in the following year. There were two engines – one by Tangyes, probably dating from the late 1890s; and the slightly larger Walker Brothers engine illustrated. The building date of c.1920 is suggested by the Serial No., but the engine is said to have been installed only a little later – in the 1920s (second-hand from a colliery). It ceased work when brickmaking was ended c.1958, but from 1960 it took over driving the pottery machinery (mixers, pumps, agitators and a conveyor) previously driven by the Tangyes, which had become inoperable. A coupled extension of the engine shaft accommodated the flywheel and two 7ft diameter belt pulleys – one (removed) used for the brickmaking machinery, and another beyond the flywheel and wall box, which drove the pottery. Its reversing gear would possibly have had little use here, also perhaps the flywheel's five rope grooves. It stopped finally in 1969, and in c.1985 the two Lancashire boilers were scrapped, but the production of stoneware (dog bowls) carried on until 1994. Both engines and at least some of the machinery remain on site, the Walker Brothers a fairly rare survival of a relatively large mill engine with piston valve, reversing gear and Pickering governor.

39. Church Gresley, Mansfield Brothers Ltd., brick manufacturers, Albert Works (SK302184)

Photo taken/Reference: 5.9.1980/232-5
Type: Horizontal single cylinder
Maker and Date: Vernon & Guest, Smethwick. c.1880/1900?
Cylinder dimensions: c.12ins x 2ft. Slide valve
Flywheel diameter: 7ft 6ins
Service: Driving brickworks machinery, making tiles, artistic pottery etc – from four direct-coupled belt pulleys of varying diameters

The works was established in 1894 by Cash, Massey & Co., and only later taken over by Horace and Fred Mansfield. The engine had, it was said, come second-hand from Sustene's flour mill, c.1920–1, a year or so after its destruction by fire. (The Sustene Ltd. cattle feed mill, in Castle Gresley, had indeed been destroyed by explosion and fire on 22nd September 1914.) It was a slide valve mill engine of the most basic type, believed to have been built by Vernon & Guest of Smethwick (this was the name said to have appeared on the plate – stolen), and dated from perhaps the final twenty years of the 19th century. The works closed in 1976, but the engine was said to have stopped work c.1970. Kilns and drying shed were still more or less intact, but demolition elsewhere had reduced most of the works to ruins. The engine remained, surrounded by much demolished brickwork, and also remaining amid the wreckage was a grinding pan by Whitaker & Co. Ltd., Accrington and a Patent Acme Victor Breaker. The site was expected soon to be cleared.

40. Derby, Derby Malt Vinegar Brewery Co. Ltd., 200 Kedleston Road (SK340375)

Photo taken/Reference: 26.5.1982/325-3
Type: Horizontal single cylinder
Maker and Date: William Abell Ltd., Derby. Probably 1906
Cylinder dimensions: c.8ins x 16ins. Slide valve
Hp: 15(max) **Rpm:** c.145 **Psi:** 35-40
Flywheel diameter: 5ft
Service: Driving air compressor and other machinery via belt-driven lineshaft

The assumed building date for the engine (1906) is derived from that of the main buildings erected when John Stretton started brewing beer on the site, having taken over from Henry Sherwin's existing brewery, which had been declared bankrupt. The change to vinegar brewing took place around 1923, when the Victoria Malt Vinegar Brewery Co. took over. The engine was still in full use in 1982. In the same room, belt-driven, was the Ingersoll Rand Company, New York horizontal (8ins x 8ins) compressor, which pumped air through the acetifiers, oxidising ethanol to produce acetic acid. On the floors above, driven via a vertical belt from the lineshaft, were a screw conveyor (for spent grain); two Steel's mashers; and a friction hoist. The brewery closed down in 1986, when the engine was removed to Derby Industrial Museum. It was of the simplest design, with single casting for engine bed and trunk guide framing, an overhung cylinder and forged crankshaft. Steam was supplied to it from a boiler working at 85 psi via a pressure reduction valve.

38

39

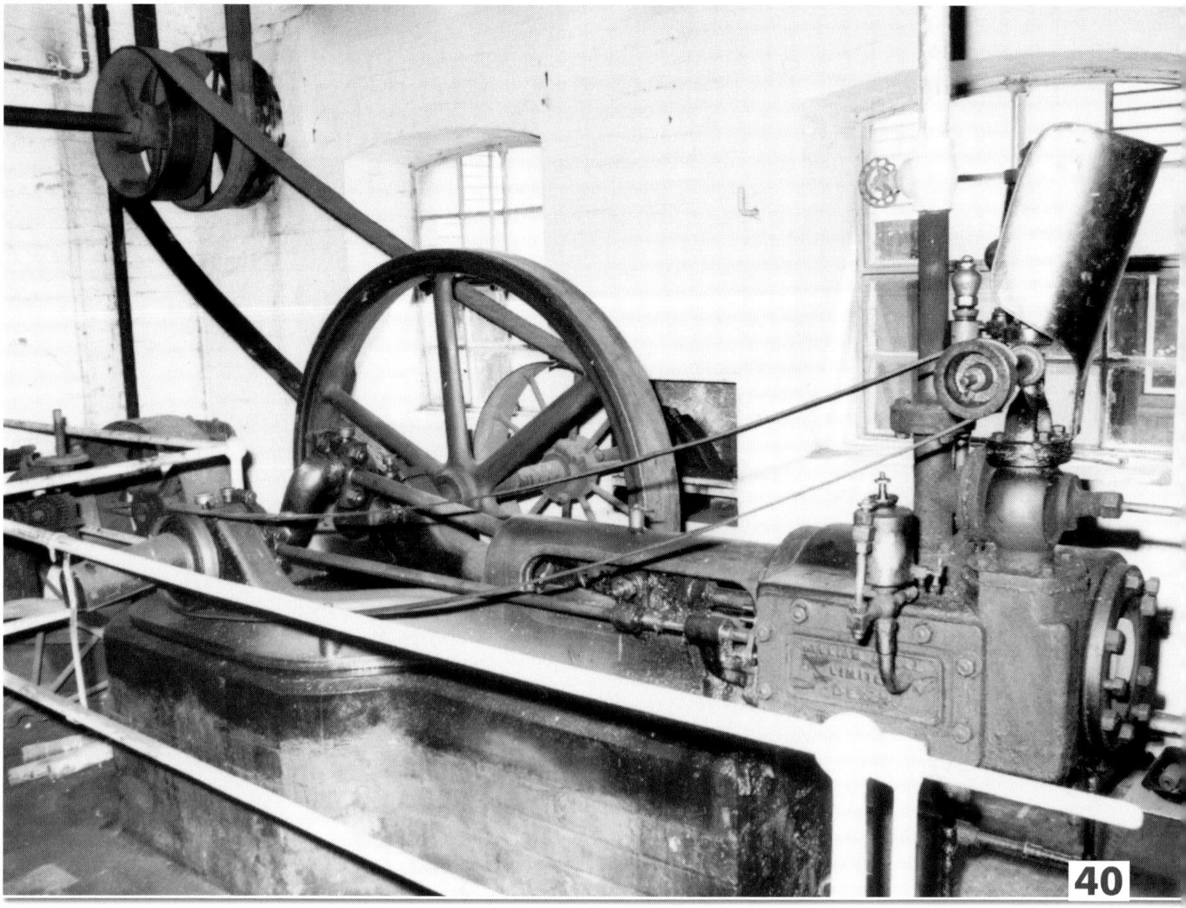

40

41. Dethick, Lea & Holloway, John Smedley Ltd., woolspinners, Lea Mills (SK318563)

Photo taken/Reference: 12.7.1988/588-31
Type: Horizontal duplex non-rotative (fire) pump
Maker and Date: Dowson, Taylor & Co. Ltd., Manchester. 1897
Cylinder dimensions: 14ins x 12ins. Slide valves
Psi: Up to 120
Service: Pumping water, originally for the sprinkler system

By around 1900, fire insurance for mills was usually dependent on the installation of an efficient sprinkler system, of which an engine such as this was a necessary part. This is an example of the standard duplex type built in large numbers by Dowson, Taylor and by Mather & Platt (into which Dowson Taylor was merged). Simple slide valves above the cylinders are used, controlled by the movement of the piston rod on the opposite side. The large air chamber on the suction box was to prevent damaging variations in the water pressure. According to the maker's specification of 24th November 1896, it was their "No. 5 Quadruple-acting fire pump" designed for "Sprinkler Fire Service", steam cylinders 14 inch, water pistons 7 inch diameter, stroke 12 inches and capable of delivering 500 gallons per minute. The cost was £150 (less a 10 per cent discount). Its use for the sprinkler system had been discontinued because it required automatic starting and a boiler always in steam, but as far as is known it is still in use for the fire hydrants.

41

42. Eckington, Renishaw Hall estate sawmill (SK439784)

Photo taken/Reference: 21.8.1994/747-7
Type: Overtype single cylinder
Maker and Date: Ruston & Hornsby Ltd., Lincoln. No. 145698. 1927
Cylinder dimensions: 8½ins x 12ins. Slide valve
Hp: 17 bhp (des.) **Rpm:** 145 (des.) **Psi:** 120 (des.)
Flywheel diameter: 4ft 9½ins
Service: Driving saw(s) by belt from direct-coupled lineshaft

The engine was taken out of service in 1979, when a new electrically-driven sawmill was erected at the estate's Hill Farm, and nothing was left of the machinery driven. An example of Ruston's semi-portable Class POG of 6 hp (nominal), it had originally gone to M.C.M. Athorpe's Dinnington Estate, in South Yorkshire, being transferred to Renishaw at about the time of the Second World War. By then (or possibly later), the engine's boiler had become unusable, and a separate vertical boiler was providing the steam. A very decorative chimney (brick with stone embellishments) rising over the sawmill was evidence of an earlier engine, and from the 1890s, power had also been provided by a small waterwheel and a turbine fed from the nearby lakes. In 1996, the engine was removed from its working site.

43. Elton, I & H. Prince & Sons Ltd., Watts Cliffe Stone Quarry, Hartshill Moor (SK222622)

Photo taken/Reference: 27.7.1976/50-5
Type: Steam derrick (single cylinder vertical engine for slewing and hoisting)
Maker and Date: Unknown
Cylinder dimensions: c.10ins x 14ins. Slide valve; Stephenson link reversing
Service: Raising building stone from quarry

This derrick was apparently in occasional use in 1976, and possibly still up to 1984, by which time ownership of the quarry had changed to Bolehill Quarries Ltd. By then, certainly, the boiler was out of use, and a compressed air line had been attached to the engine. Some time between 1985 and 1988, it was scrapped – a sad loss of a rare example. Slewing was achieved by means of two facing bevel gears on the crankshaft, either of which could be clutched on to a bevel on a vertical shaft which at its foot was double reduction geared to a toothed ring fixed to a large foundation block. Hoisting was by single reduction gearing to the drum shaft, or by either of two ratios of double reduction.

44. Killamarsh, Westthorpe Colliery (SK453797)

Photo taken/Reference: 2.7.1987/557-16
Type: Horizontal twin cylinder winding engine
Maker and Date: Robey & Co. Ltd., Lincoln. Nos.40223–4. 1924
Cylinder dimensions: 22ins x 40ins. Drop valves
Psi: 150
Service: Winding coal from shaft, depth 151 yards. Semi-conical drum 9ft – 12ft diameter

Robey's Class E coupled winding engine, this was built for the single shaft of J. & G. Wells Ltd.'s Westthorpe Colliery (there was an underground linkage to Holbrook Colliery). It was used initially mainly for man winding, but from c.1946–7 changed to coal winding (for which the maximum speed was 45ft/sec). The valves were driven from a skew gear driven layshaft, with admission and exhaust valves being operated by a single eccentric at each end of the cylinder. Second eccentrics drove the trip gear controlled by the Richardson governor. The usual Robey reversing gear was fitted. By the mid 1970s, the governor had been disconnected, and by 1979, the trip gear mechanism had been largely removed. By then, also, although the engine had a cylindro-conical drum, it was working "using only the flat". This was one of the last collieries to retain steam, until production stopped in 1984. The engine lingered on until c.1998, while various options were being considered – firstly, continued operation in conjunction with Highmoor Colliery and then an in situ preservation scheme. It is now being preserved at Pleasley Colliery.

42

43

45. Sandiacre, Sandiacre Mill Company Ltd., Springfield Mill (SK480365)

Photo taken/Reference: 19.8.1984/456-33
Type: Inverted vertical single cylinder rotative (fire) pump
Maker and Date: Merryweather & Sons Ltd., London. No. 1960. Possibly c.1888
Cylinder dimensions: c.12ins x 8ins. Slide valve
Flywheel diameter: 38½ins
Service: Mill fire pump. Pump, double-acting, c.8ins diameter

This was an example of one of Merryweather's patent design of fire pumps, the only one known still remaining in situ into the 1980s. The mill was erected in 1888 for lace manufacture by Terah Hooley Limited, and the pump may be of the same date. Having been taken over by the Sandiacre Mill Company in 1919 and let out to various companies, the mill when visited was still in use for lace manufacture. Semi-derelict, the pump was believed to have been out of use for many years, and was removed to Papplewick Pumping Station about three years later. An interesting design, the pump was direct-driven from the steam cylinder by a single rod, with a crankshaft and flywheel to one side, controlling the stroke and operating the slide valve spindle from a disc crank. The piston/pump rod is connected by a fixed double plate link to a guided crosshead from which the connecting rod rises to the crankshaft.

46. Wingerworth, National Coal Products Ltd., Avenue Coking Works (SK391677)

Photo taken/Reference: 26.1.1990/644-22
Type: Inverted vertical compound compressor (enclosed)
Maker and Date: George Waller & Son Ltd., Stroud. No. 3057. 1956
Cylinder dimensions: 20ins/28ins x 12ins. Piston valves
Hp: 450 (max des.) **Rpm:** 150 (normal max – 300 des. max) **Psi:** 80
Flywheel diameter: c.4ft 3ins
Service: Raising pressure of coal gas produced by coke ovens

When built by the National Coal Board's Coal Products Division in 1956, this was its largest coking plant, believed to be the most modern in Great Britain. There were two batteries, each of 53 coke ovens. The Gas Booster House contained two steam-driven compressors and two electric motor driven sets, all of equal maximum design capacity: 6 million cubic ft of gas per day to be delivered at a pressure of up to 25 psi. Each of the steam compressors had two single-stage compressor cylinders, 28½ins in diameter, in tandem above the steam cylinders. There was a "piston type steam governor to control the outlet (gas) pressure". The usual practice then current was to operate one motor-driven set (at a constant 300 rpm) with the variation in requirements being taken up by a steam set – varying in speed but usually compressing around 2 million cubic ft per day. (The outlet pressure actually observed was around 15–16 psi.) Only just over two years later (in September 1992) the plant was shut down, demolition began in May 1999, and the steam compressors were reportedly sold for export. (The second engine, George Waller No. 3058, under repair, is behind to the right in the illustration.)

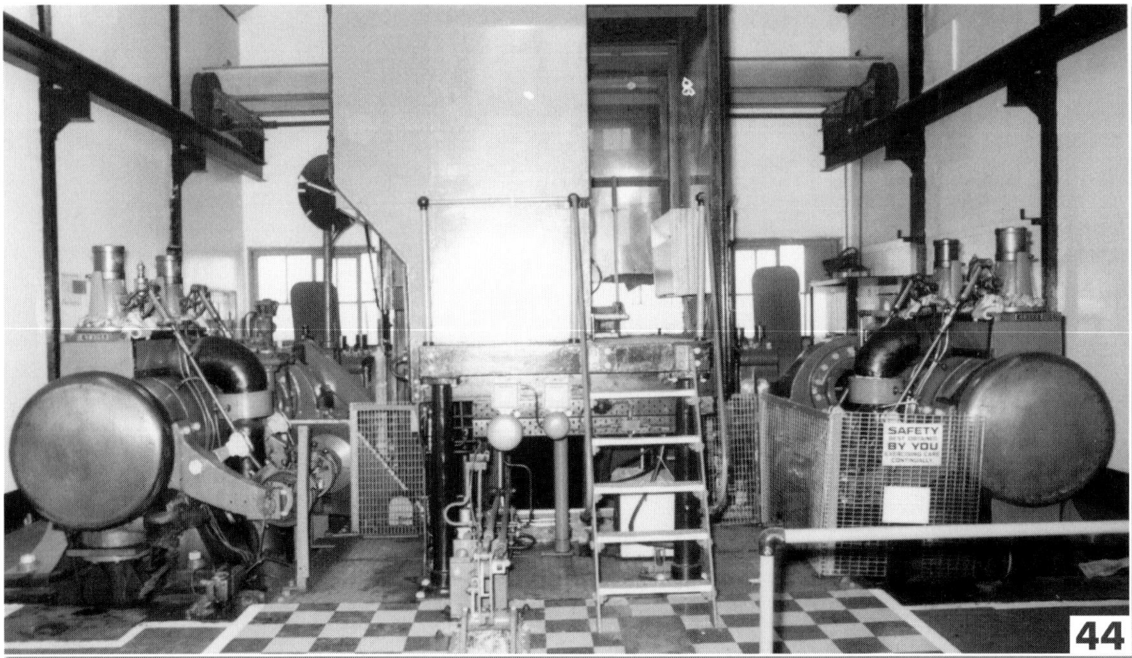

44

45

46

47. Woodville, John Knowles & Co. (Wooden Box) Ltd., fireclay goods manufacturers, Mount Pleasant Works (SK312184)

Photo taken/Reference: 27.4.1973/4639
Type: Horizontal single cylinder
Maker and Date: Buxton & Thornley, Burton-on-Trent. c.1880s–1890s
Cylinder dimensions: 16ins x 30ins. Slide valve
Hp: 82 ihp? **Rpm:** 80 **Psi:** 80
Flywheel diameter: 9ft
Service: Driving brick mill machinery – pug mill and brick machines. Belt drive from 5ft diameter pulley to lineshaft

The Mount Pleasant Works was started by John Knowles in 1849 for the manufacture of refractory (fireclay) goods and stoneware sanitary pipes. As late as the 1960s, there were apparently at least five engines on the site, all horizontal single cylinder. These included two claypit winding engines by Thornewill & Warham, removed for preservation in 1966, and three Buxton & Thornley engines which survived longer; an engine in the fitting shop, removed for preservation, perhaps around 1970; and the Pipe Mill and Brick Mill engines, of about the same size, dating from around the 1890s, and probably stopping work in the latter half of 1971. The Pipe Mill engine was scrapped in 1973, with the Brick Mill engine, shown here, being removed to the Snibston Discovery Centre as late as 1997. By the end of the 1980s, the works was contracting in size. The kilns and the remainder of what had been a forest of chimneys were demolished in the mid 1980s, and in 1997 the remainder of the old works was cleared.

48. Youlgreave, Long Rake Spar Co. Ltd., Long Rake Mine (SK187642)

Photo taken/Reference: 27.7.1976/49-32
Type: Horizontal twin cylinder geared hoist
Maker and Date: John Wood & Sons Ltd., Wigan. Nos. 3592–3. 1925
Cylinder dimensions: 6ins x 12ins. Slide valves; Stephenson reversing links
Psi: c.80
Service: Winding mineral and men from a depth of 267ft. The small cage could take 3 men or a c.½ton capacity tub.

This was a calcite mine which dated from the early 1870s, but the 1920s appear to have seen an increase in activity linked to the formation of a new company: the Long Rake Spar Company. The shaft was sunk to a depth of 267ft, using this engine, purchased new from John Wood in 1925, and a vertical boiler. The mine eventually had two lower levels, but these were reached by an incline. As the mine's sole winding engine, it seemed small, but it appears to have given good service over more than 50 years. A single drum, 42 inches in diameter, was divided by a central brake path, and driven by chevron gearing (single reduction c.4.8:1). The mine closed in 1980 – perhaps partly because the boiler, although by then a replacement, a vertical by Cradley Boiler Co. Ltd., Cradley Heath (No. 20104 of 1934), needed expensive repairs. More important was the depressed price of calcite and diminished reserves. However, the surface dressing plant continued to operate, and with the shaft reportedly in use for pumping from an adjacent mine, the engine is believed for a while to have been used for occasional inspections (probably using compressed air). It was removed c.2003 for possible preservation.

DORSET

49. Corfe Mullen, Corfe Mullen Pumping Station (SY974982)

Photo taken/Reference: 15.4.1981/255-6
Type: Inverted vertical triple expansion pumping engine
Maker and Date: Hathorn, Davey & Co. Ltd., Leeds. No. 7992. 1932
Cylinder dimensions: 17ins/29ins/44ins x 42ins. Drop valves
Hp: 250 bhp **Rpm:** 25 **Psi:** 180
Flywheel diameter: 14ft
Service: Pumping up to 2.8 million gallons/day to reservoir for water supply of Poole

Poole Corporation Water Works had two similar triple expansion engines by this maker, a smaller one having been installed in 1925 but already scrapped. Unlike the earlier engine, which had Corliss valves, this was Hathorn, Davey's later type with drop valves, although still with a cylinder level camshaft and Craig's valve gear. The well pumps (18½ins x 42ins buckets) were driven via bell cranks from a two-throw extension of the crankshaft; and three ram pumps (15ins x 42ins) were driven from the engine crossheads. Around 1986/7 the engine was removed for preservation, but 20 years later was at Preston, near Canterbury, in pieces and still without a permanent home.

47

48

49

DURHAM

50. Evenwood & Barony, Randolph Coke & Chemical Co. Ltd. (NZ157248)

Photo taken/Reference: 14.5.1983/389-24
Type: Inverted vertical twin cylinder (enclosed)
Maker and Date: W. Sisson & Co. Ltd., Gloucester. No. 5930. c.1957
Cylinder dimensions: 15ins x 9ins. Piston valves
Hp: 215 bhp **Rpm:** Variable (391 – des. max) **Psi:** 90 (des.)
Flywheel diameter: 4ft 6ins
Service: Direct-coupled to Roots-type exhauster by G. Waller & Son Ltd., Stroud. No. 3404. 24ins x 20ins. (Gas Booster No. 2.)

Still to be seen working here until the plant stopped production on 15th March 1984 was a splendid array of six high-speed enclosed engines – all built by Sisson, a company still making reciprocating steam engines into the 1960s. Driving exhausters and installed by Sadler & Co. Ltd., then the owners of the plant, to deal with the coal gas produced from the company's coke ovens, the first three to be put in were linked to the erection of 15 ovens in about 1948, with the others following. In the Exhauster House were three single cylinder engines – one working at a constant speed, one as a standby, and a larger one, varying in speed to take up the load. In the Booster House was this Sisson's type TDC (twin double crank) engine, which had a piston valve for each cylinder and a shaft governor controlling automatic expansion gear. Here also, as the largest engine, it ran at a speed which varied in accordance with the fluctuating load, while the other two in the booster house – single cylinder engines of 1946 and 1949 – worked at a constant 110–120 rpm.

51. Hartlepool, George Clark & NEM Ltd. (formerly Richardsons Westgarth Ltd.), Ferry Road (NZ522338)

Photo taken/Reference: 27.3.1974/8-26
Type: Stationary wharf crane
Maker and Date: Cowans, Sheldon & Co. Ltd., Carlisle. No. 2661. 1904
Cylinder dimensions: Three twin cylinder engines with slide valves and link reversing gear:
(1) Slewing engine. c.10ins x 12ins
(2) Hoist engine. c.10ins x 14ins
(3) Smaller hoist engine (outside crane)
Service: Dockside crane, maximum capacity 75 tons

Cowans, Sheldon had been capable of producing a wide variety of engineering products, including most significantly railway turntables and railway cranes. Other cranes, such as the Fairbairn type seen here, were also made in large numbers. Unfortunately, very little information about this crane has come to hand, although it is believed that Hartlepool Museum has copies of drawings etc. By about 1972, George Clark & NEM's lease on the crane's dockside site had expired, and the crane became the property of the Tees & Hartlepool Port Authority. Its foundations were said to be in danger of falling into the harbour and despite attempts at preservation, it was demolished in May 1975.

52. Hunstanworth, Sikehead Lead Mine – Ellen's Shaft (NY955464)

Photo taken/Reference: 12.7.1980/229-31
Type: Horizontal twin cylinder winch
Maker and Date: J.C. Petersens Mekaniska Verkstads A.B., Trelleborg, Sweden
Cylinder dimensions: 7ins x 10ins. Slide valves; Stephenson reversing links
Service: Winding lead ore or fluorspar from Ellen's Shaft

This small winch is assumed to have come off a scrapped ship, and was installed with a vertical boiler by Hunstanworth Mines Ltd., probably c.1920, when the mine was re-opened to produce lead and fluorspar. It presumably stopped working in 1931 when the mine was closed, but there was further production in the period 1945–1983 and it is possible that the winch was brought back into use. The drum was driven via 6:1 reduction to a gear ring attached to the drum or alternatively in two stages with an extra intervening (2.5:1) reduction. The winch was removed or scrapped around 1989.

50

51

52

53. Sunderland, Armstrong, Addison & Co. Ltd., timber creosoters, Pier View, North Dock (NZ408585)

Photo taken/Reference: 10.7.1980/227-34
Type: Steam derrick crane
Maker and Date: Butters Bros. & Co. Ltd., Glasgow
Cylinder dimensions: Two vertical cylinders, c.6ins x 10ins. Slide valves; Stephenson links
Psi: c.80
Service: Moving timber

Established in 1853 and supplying preserved timber and wooden fencing on a large scale, the Armstrong, Addison timber yard was remarkable for the amount of steam machinery continuing in use until around the early 1980s. This included two winches for drawing trucks of timber into and from a creosoting cylinder; three travelling cranes; and two derricks built by Butters Bros. (A Holman Bros. air compressing and vacuum pumping engine of 1914 used in the creosoting process had been removed for preservation in 1978.) One of the derricks was built in 1938. The other, the one illustrated, is the larger of the two (with a 120ft jib), and was said to have been bought for £250 from the River Wear Commissioners c.1966, but its building date seems to be unknown. There is a single pair of vertical cylinders for hoisting, derricking and slewing – the latter via facing bevel gears.

ESSEX

54. Coggeshall, G.P. Swinborne & Co. Ltd., isinglass manufacturers (TL841224)

Photo taken/Reference: 15.11.1976/53-18
Type: Horizontal single cylinder
Maker and Date: W. Pope & Son, Slough. 1886
Cylinder dimensions: c.12ins x 28ins. Slide valve; Stephenson link reverse
Hp: 30 (nom)
Flywheel diameter: 9ft
Service: Belt drive from 4ft diameter pulley to lineshaft, from which belts drive machinery

The Swinborne works was established in 1848 to make gelatine and isinglass – particularly the latter, used for finings in beer production, in later years. But the engine was possibly installed in connection with an expansion around 1886, the date included in the strip of ornamental tiles round the base of the chimney stack. With a cast iron bed supported on a brick base and a separate tall brick pier for the bearing on the far side of the flywheel, this was an interesting engine, quite likely to be the only one by this obscure maker to survive. Features included the wood strips on the cylinder; a large (35 inch diameter) disc crank; a stay between the ends of the guide bars, looped up to allow for the connecting rod; and a slide valve driven via a reversing link, although the reason for this was not identifiable. The engine had ceased regular work possibly in about 1947, and was removed for preservation shortly after the photograph was taken. The works closed altogether in the 1990s.

53

54

55. Felsted, T.D. Ridley & Sons Ltd., brewery, Hartford End (TL688175)

Photo taken/Reference: 12.3.1979/144-30
Type: Horizontal single cylinder
Maker and Date: Tangyes, Birmingham. Installed 1890s?
Cylinder dimensions: 7½ins x 12ins. Slide valve
Flywheel diameter: 4ft 10ins
Service: Drove brewery machinery, presumably from 22in diameter belt pulley

This small and picturesque country brewery was established by Thomas Dixon Ridley in 1843, using the site of an existing cornmill. The engine was housed in what was effectively a small cupboard (making photography difficult), in a later extension, understood to have been built in the 1890s. Assumed to be by Tangyes (it is fitted with their "Soho" type governor), it has the appearance of being possibly older even than this. Design features include the flat bed, supporting the two main bearings and bent crankshaft between, and two bar crosshead guide – a single bar on each side with guide surfaces above and below. The engine appeared to have been out of use for some time and was removed c.1983–4 (it is now at the Museum of Power, Langford). Sadly, the brewery was closed in 2005, following a takeover by Greene King.

56. Ilford, South Essex Water Works Company, Roding Pumping Station (TQ416892)

Photo taken/Reference: 26.7.1979/170-37
Type: Inverted vertical triple expansion pumping engine. No. 2 Engine
Maker and Date: Hathorn, Davey & Co. Ltd., Leeds. No. 6361. c.1909
Cylinder dimensions: 15ins/25ins/40ins x 3ft. Corliss valves (Craig's valve gear)
Hp: c.300 **Rpm:** Up to 40? **Psi:** 150
Flywheel diameter: 15ft
Service: Two Ashley bucket well pumps (20½ins x 42ins), and three single-acting ram pumps (10½ins x 36ins) pumping 2 m.g.d. to reservoir at Hainault

The Company installed three almost identical engines from Hathorn, Davey, one each at Ilford and Roding Stations in 1905, and a third, illustrated here, at Roding. (The space provided for a second engine at Ilford was never needed!) Corliss valves were operated from a cylinder level eccentric/cam shaft, with a modified Proell governor controlling the HP trip gear. The two engines at Roding were arranged end-on, with well pumps between – two for each engine, driven from a pump crankshaft geared down 2.75:1, a horizontal connecting rod, and a pair of quadrants. The high-level ram pumps were driven from the engine crossheads. The No. 1 Engine was largely scrapped following the conversion to electric pumping 1957–61, but the No. 2 Engine lingered on until 1987. There were three Economic boilers by Robert Taylor & Sons, Marsden, 1905.

57. Langford, Essex Water Company, Langford Pumping Station (TL835090)

Photo taken/Reference: 7.6.1980/221-15
Type: Inverted vertical triple expansion pumping engine
Maker and Date: Lilleshall Co. Ltd., Oakengates. No. 282. 1931
Cylinder dimensions: 20ins/35ins/56ins x 42ins. Double-beat drop valves
Hp: 350 bhp **Rpm:** 20–23 (max 33) **Psi:** 210
Flywheel diameter: 14ft
Service: Water supply to Southend. Three low lift ram pumps, 18¾ins x 36ins stroke (head 30ft). Three high level ram pumps, 16¾ins x 42ins stroke (head 240–300ft). Capacity (at 30 rpm) 4 m.g.d.

This was the sole survivor (now preserved) of three identical engines, built 1927–31 for the Southend Waterworks Company. Three Babcock & Wilcox water tube boilers supplied the steam, but these, with the other two engines, were scrapped after they stopped work in 1963. Double-beat drop valves driven by a high-level eccentric shaft were fitted, with a "Bee" governor (William Eastwood's patent) to control the HP inlet valve trip gear. The raw water was drawn from the rivers Chelmer, Ter and Blackwater, gravitating into a storage reservoir, from which after sedimentation it was pumped to the Treatment Plant by the low lift pumps, driven by a 3-throw extension of the crankshaft. The high level pumps, driven from the engine crossheads, pumped the treated water to reservoirs at Oakwood, c.14miles distant.

55

56

58

58. Mistley, EDME Ltd., malt extract works (TM118317)

Photo taken/Reference: 13.6.1975/31-17
Type: Horizontal single cylinder
Maker and Date: Marshall Sons & Co. Ltd., Gainsborough. No. 65019. 1914
Cylinder dimensions: c.10ins x 14ins. Slide valve
Psi: 120?
Flywheel diameter: 5ft
Service: Driving three-throw well pumps by George Adlam & Sons Ltd., Bristol

This was the works of the English Diastatic Malt Extract Co. Ltd.The engine was a Marshall Class MP engine – with slide valve Pickering governor, overhung cylinder and bent crankshaft – one of the group of M engines introduced around 1895. It was connected to the separate pump assembly and possibly to other machinery on the floor above by means of belts to and from a ceiling level lineshaft, the final drive to the pump shaft being by reduction gearing.The engine was only being run one day a week in 1971, and it was stopped completely shortly after. In 1982 it was removed to the Forncett Industrial Steam Museum, Norfolk.

59. Orsett, Baker Street Mill (TQ633813)

Photo taken/Reference: 29.6.1975/31-30
Type: Single cylinder rotative beam
Maker and Date: Thomas Middleton, Southwark. Possibly c.1840
Cylinder dimensions: 9ins x 24ins. D-slide valve
Flywheel diameter: 8ft
Service: Driving two pairs of stones in cornmill

The windmill on this site was recorded as existing in 1796, and the completely independent steam mill is believed to have been added around 1840 – either by James Woollings or by his successor William Woollings.The free-standing A-frame engine was in a lean-to and directly drove a bevel gear in the main building, from which power was taken via a vertical shaft to the stones above. Steam milling had probably ceased by 1910 (although a replacement locomotive-type boiler had been installed in 1906) and the windmill ceased work c.1914–15. However, much of all this remained for many years, the picturesquely derelict steam mill retaining its chimney and incomplete engine. Also of interest was a wall-mounted vertical boiler feed pump (largely intact), driven by belt from the engine.Today, the engine remains and millwork are incorporated into a dwelling house, built on the site in 1982. (As will be seen from the illustration, the engine had lost a number of its major parts, including beam, connecting rod, eccentric rod and governor.)

59

57

60. Roydon, A. Dyke & Sons, Netherhall Nursery (TL400080)

Photo taken/Reference: 9.11.1997/842-19
Type: Inverted vertical compound (enclosed)
Maker and Date: Peter Brotherhood Ltd., Peterborough. No. 43237. 1930/1
Cylinder dimensions: 9½ins/15ins x 8ins. Piston valves
Hp: c.90 bhp **Rpm:** 500 **Psi:** 120
Flywheel diameter: 3ft 3ins
Service: Electricity generation. Direct-coupled to 60 kW alternator, Crompton Parkinson Ltd., Chelmsford & Guiseley No. 157334 This was a fairly rare surviving example of an engine made by a company more usually known for its three cylinder radial engines. Seen just after the nursery had been sold – possibly for housing development – it was only a few weeks later moved to the Cambridge Museum of Technology, Cheddar's Lane. With little unusual about its design, it had two piston valves, between the cylinders, worked by a single eccentric. The crankshaft governor was linked only to a throttle valve. It had been first used in the Ideal Laundry, Chatham. By 1944 and until possibly c.1957 it was owned by the Leys Steam Laundry, Cambridge, for whom it was reconditioned by the builder. Finally, in 1972 it was purchased by Dykes, apparently from a North Weald nursery (where, however, it had never been installed), continuing to work until around the mid 1980s.

61. Tiptree, Wilkin & Sons Ltd., jam manufacturers (TL899156)

Photo taken/Reference: 3.5.1975/30-16
Type: Horizontal single cylinder
Maker and Date: Marshall Sons & Co. Ltd., Gainsborough. No. 82164. 1927
Cylinder dimensions: 8ins x 20ins. Drop valves
Hp: 65 **Psi:** 80
Flywheel diameter: 7ft 6ins
Service: Driving 40 kW generator from 6ft diameter belt pulley

This factory was established by Arthur Wilkin in 1885 to make use of the product of his fruit farms at Tiptree. Compact and solidly built, this was a small, late example of Marshall's L Class engine, which unfortunately had ceased work around 1970. By 1975, when the engine was photographed, the generator had gone, and the engine itself was to be shortly afterwards removed to Strumpshaw Hall Steam Museum (Norfolk). All the valves and the chain-driven Proell governor were operated via a layshaft driven by mitre wheels from the crankshaft. The double-beat steam valves were driven by eccentrics and have Marshall's patent trip gear controlled by the governor. In larger engines of this type, the exhaust valves, below the cylinder, were also operated by eccentrics. But in the smaller engines, as here, the valve rods have attached levers and rollers which are operated by cams.

62. Walthamstow, Low Hall Farm Sewage Works, Markhouse Avenue (TQ363883)

Photo taken/Reference: 12.7.1978/118-15
Type: Two horizontal single cylinder engines (coupled)
Maker and Date: Marshall Sons & Co. Ltd., Gainsborough. Nos. 27384-5. 1896–7
Cylinder dimensions: 12ins x 24ins. Slide valve and expansion valve
Hp: 60 (each) **Psi:** 120
Flywheel diameter: 9ft
Service: Driving two centrifugal pumps raising effluent from low level sewer – originally to filtration and settling tanks, but from 1928 into L.C.C. sewage system. The average volume of sewage pumped per day: 300,000 gallons. Average lift 20ft

The first steam pumps on the site were by Hayward Tyler & Co., and erected in 1885. These two engines – Marshall's Class C, fitted with Hartnell's patent automatic expansion gear – were installed new as replacements, together with a Cornish boiler. In 1905 two Lancashire boilers with Meldrum's refuse destructor furnaces were put in. The pumps were driven via a lineshaft connected to the engines by belt from a common flywheel. All this was superseded by electric pumps when the boiler plant was condemned, probably in 1962, but the engines are now being preserved in situ.

60

61

62

63. West Ham, Tate & Lyle Refineries Ltd., Thames Refinery, Silvertown (TQ422800)

Photo taken/Reference: 8.10.1974/19-29
Type: Horizontal single cylinder rotative vacuum pumps
Maker and Date: Harvey Engineering Co. Ltd., Glasgow. Nos. 3052 & 3053. 1920
Cylinder dimensions: 24ins x 3ft. Slide valve
Rpm: c.30 (originally 60) **Psi:** 55–60 (reduced from 100)
Flywheel diameter: 14ft
Service: Dry air vacuum pumps providing 25–27in vacuum for sugar evaporation pans

These engines were ordered by Henry Tate & Sons from the Harvey Engineering Company (Order No. 9716), an important manufacturer of sugar machinery, in 1919. By the time the engines started work (in January 1923 according to the engine plates) the Tate and Lyle companies had amalgamated. They were heavily built slide valve engines, with 4-bar guides and Pickering type governors, each air cylinder (40 inches in diameter), in tandem behind the steam cylinders, having its own slide valve and eccentric rod. They were taking turbine exhaust steam at 60 psi and exhausting to process at about 13 psi. They remained in use until 1976, then on standby until 1979, after which No. 3052 (on the right in the photograph) was removed to the National Museum of Scotland and the other one scrapped.

64. West Ham, West Ham Sewage Pumping Station (TQ389833)

Photo taken/Reference: 3.10.1973/5-21
Type: Inverted vertical compound
Maker and Date: John Cochrane & Co., Barrhead. 1900
Cylinder dimensions: 20ins/33ins x 2ft. Piston valve (HP); slide valve with Meyer expansion valve (LP)
Hp: 274 **Rpm:** 133 (des.) **Psi:** 100 (des.)
Flywheel diameter: 7ft 9½ins
Service: One of three engines direct-coupled to centrifugal pumps, delivering surplus storm water and sewage into the River Thames. Capacity (one engine) 266,500 cubic ft/hour (head 25ft)

Three inverted vertical compound engines, driving centrifugal pumps, were erected for the County Borough of West Ham for pumping storm water – two by Cochrane; and a smaller one of possibly German make, second-hand from Canning Town Pumping Station. (The main pumping engines of 1895–1900 are illustrated in *Stationary Steam Engines of Great Britain*, Volume 9 – see No.54.) The Cochrane engines were identical except for being opposite hand, the HP cylinder being fitted with a piston valve of which the travel and cut-off were controlled by a Hartnell governor and variable link. There were surface condensers below, each with an air pump operated by a disc crank at the HP end. The "German" engine had gone before this photograph was taken – apparently for preservation, although its present location seems to be something of a mystery; and the Cochrane engines were scrapped soon after.

GLOUCESTERSHIRE

65. Bristol, Port of Bristol Authority Workshops, Underfall Yard (ST572721)

Photo taken/Reference: 25.8.1976/52-16
Type: Horizontal twin cylinder
Maker and Date: Tangyes Ltd., Birmingham. Nos. 5123-4. 1885
Cylinder dimensions: Probably 8ins x 16ins. Slide valves
Hp: 40–50 bhp **Rpm:** 60 **Psi:** 40–50
Flywheel diameter: 59ins
Service: Driving workshop machinery

The engine and much of the machinery date from around 1885, when there was a considerable expansion of the dockyard maintenance workshops. All remains much as when installed and is a valuable survival of a complete 19th-century workshop. The engine consisted essentially of two opposite handed Tangyes E size engines with a common crankshaft and flywheel between. Drive was taken by flat belt from the flywheel to a high level lineshaft running along one side of the length of the machine shop, and from this there was a belt connection across to a second shaft running along the other side. Speed control was by a Tangyes throttle governor. Steam was provided by a Cornish boiler with a wall-mounted banjo boiler feed pump by A.G. Mumford Ltd.

63

64

65

66. Bromsberrow, Bromsberrow Pumping Station (SO738333)

Photo taken/Reference: 3.7.1991/672-8
Type: Horizontal side-by-side compound pumping engine
Maker and Date: Tangyes Ltd., Birmingham. 1928
Cylinder dimensions: c.14ins/c.24ins x 2ft. Slide valves; Meyer expansion valves
Hp: 150 **Rpm:** 56 **Psi:** 100
Flywheel diameter: 9ft
Service: Water supply to Malvern. Capacity 30,000 gallons/hour (total head 600ft)

This station was established by Malvern District Council in 1906, but the first two engines had been scrapped. Still remaining, although out of use since 1970, was this engine, which pumped from a third 200ft borehole sunk in 1927. The valves were outside, both cylinders having Meyer expansion valves; driven by a gear wheel next to the LP crank was a Pickering type governor linked to a "knock-off" inlet valve standing on the HP valve chest. All the pumps had been removed: a well pump driven by gearing from the flywheel end of the crankshaft, and at the other end three horizontal ram pumps driven from a coupled extension crankshaft. At the outer end of this there had been a crank for an air pump and condenser below (all gone before 1977).

67. Christchurch, Arles Level coalpit (SO572134)

Photo taken/Reference: 5.4.1992/679-29
Type: Horizontal twin cylinder haulage
Maker and Date: N.V. Coevorder Machinefabriek. Installed 1932?
Cylinder dimensions: c.7ins x 12ins. Slide valves; Stephenson link reverse
Service: Rope haulage of coal tubs up pit incline

Arles Level was a small Forest of Dean coal mine operated by the Gwilliam Brothers from September 1913 to probably around 1950. This Dutch-made engine was one of two which, with a small vertical boiler and considerable quantities of rubbish, still remained in a corrugated iron shed in the garden of Arles House. Possibly of marine origin, it was said to have been "supplied by Adams of Newport in 1932". The drum (c.15ins diameter and 29ins width) was driven by (c.5:1) reduction gearing. Provision was made in the frame for an intermediate shaft (not present) for double reduction. Unfortunately, the dog clutch for disengaging the gear drive for the free return run to the mine for some reason proved to be unsatisfactory, and the second similar engine by Clarke, Chapman & Co. was installed as a replacement a few years later. Both engines have recently been removed to the nearby Clearwell Cavern for preservation.

HAMPSHIRE

68/69. Fleet, Grab Dredger *Perseverance*, on Basingstoke Canal

Photo taken/Reference: 6.3.1993/710-1 & 710-11
Type: Floating grab dredger – Horizontal twin cylinder engine
Maker and Date: Grafton & Co., Bedford. No. 2473. 1934
Cylinder dimensions: 7ins x 10ins. Slide valves; Stephenson link reversing
Service: Canal dredging

Seen here probably a week before its retirement, this dredger's final task, from 1974 to 1993, was to assist with the Surrey and Hampshire Canal Society's restoration of the Basingstoke Canal, after having been acquired from the Kennet & Avon Canal. A pontoon dredger, *Perseverance* consists of an unpowered narrow boat which was supplied to the Grand Union Canal Company in 1934, with a Grafton steam grab crane mounted on one end, and at the other end below deck a Marshall locomotive type boiler of 1947, second hand from a Luton laundry. Detachable pontoons were added to the sides to give the vessel stability, and while in use it was positioned and held by hand winches and wire ropes to the bank. A single horizontal engine with side-by-side cylinders and outside valves provided the power for hoisting, derricking and slewing through a complex of gearing. A direct-acting $10^{1/2}$in diameter steam cylinder mounted on the lower jib opened and closed the 3 ton grab. All this operation required considerable skill on the part of the driver, and surrounded by much escaping steam, was a memorable sight. Soon afterwards, it was taken to the Ellesmere Port Canal Museum.

66

67

68

69

70. Portsmouth, H.M. Naval Dockyard – No. 2 Slipway (SU628009)

Photo taken/Reference: 13.8.1979/174-31
Type: Inverted vertical twin cylinder haulage winch
Maker and Date: Cowans, Sheldon & Co. Ltd., Carlisle. No. 194. 1896
Cylinder dimensions: c.9ins x 12ins. Slide valves; Stephenson link reversing
Service: Slipway haulage

The Cowans, Sheldon company is known principally for its cranes, particularly railway steam cranes, but also produced other machinery, of which this is an interesting example. It was of very substantial construction, the drum being driven via three pairs of reduction gears and two intermediate shafts, the first of which carried a brake wheel and a clutch enabling the drum to be decoupled from the engine. Originally connected to the dockyard steam mains, it was last worked by compressed air, although out of use when photographed. Cowans, Sheldon were also responsible for some of the twin cylinder engine driven capstans in the dockyard, two of which, built in 1910, were used for operating the No. 5 Dock gates. The winch was removed in 1982 and is now preserved at Chatham Dockyard.

71. Sarisbury, Redland Bricks Ltd., Bursledon Brickworks, Lower Swanwick (SU500099)

Photo taken/Reference: 9.9.1994/753-19
Type: Horizontal single cylinder
Maker and Date: John Wood & Co., Wigan? Probably c.1885
Cylinder dimensions: 16ins x 30ins. Slide valve
Rpm: 80 **Psi:** 100
Flywheel diameter: 8ft
Service: Direct-coupled to Bennett & Sayer, Derby brick machine (for wire-cut bricks)

It is believed that this engine was first installed by Hooper & Ashby at their Red Brick Works, Chandler's Ford around 1885, being moved when operations were transferred to Lower Swanwick in 1897. By around 1903, the works were operated by the Bursledon Brick Company, and it was then that the engine was modified by Tangyes and the Pickering type governor fitted. The works finally ended up in the hands of Redland Bricks from 1959 until closure in 1972. A second engine, definitely by John Wood, was removed to Wendron Forge for preservation c.1978–9, but by 1993 the works had become The Centre for the Conservation of the Built Environment, and the earlier engine was retained on site. The photograph was taken just after the engine had been nicely restored by Ian Clark.

HERTFORDSHIRE

72. Bourne End, J.W. Ward & Son Ltd., sawmills, wood turners etc. (TL019061)

Photo taken/Reference: 22.5.1981/276-28
Type: Horizontal tandem compound
Maker and Date: Davey, Paxman & Co. Ltd., Colchester. No. 22351. 1926
Cylinder dimensions: 14ins/23ins x 2ft. Drop valves
Hp: 250 bhp (des.) **Rpm:** 150 **Psi:** c.90–110
Flywheel diameter: 8ft
Service: Driving General Electric, Schenectady, N.Y. 150 kW alternator by 4 ropes from engine flywheel (only 4 of 12 grooves in use). The speciality product was brushmakers' woodwork

Davey, Paxman built around 130 engines with Lentz positive valve gear, about 100 of which were horizontal tandem compounds. This was one of two such engines which remained in use until 1980 (for the other see *Stationary Steam Engines of Great Britain*, Volume 6, No. 39). Research by the late Graham Smart established that it was built for the Aylesford Pottery Co., in Kent, but the date of installation at Bourne End has yet to be ascertained – it was just possibly in the early 1930s following a cutback in the Pottery's product range. As usual with this type, HP cut-off was by the Paxman-Lentz "Inertia" type (layshaft) governor controlling the inlet valve eccentric rods. There was no condenser, exhaust being to atmosphere or heating. The step up to the alternator speed of 1, 000 rpm was in two stages via an intermediate shaft. A declining "Economic" type boiler, also by Davey, Paxman, stopped the engine, and shortly afterwards, in 1981, the sawmill ceased operations. The engine was removed for private preservation.

73. Boxmoor, J.W. Ward & Son Ltd., Fosters Sawmill

Photo taken/Reference: 22.2.1969/3281
Type: Horizontal single cylinder
Maker and Date: Vernon & Guest, Smethwick. c.1880?
Cylinder dimensions: c.20ins x 40ins? Slide valve; Meyer expansion valve
Rpm: 55?
Flywheel diameter: 12ft
Service: Belt drive to saws from pulley

This engine was said to have been used in a London works until 1913, and installed here in 1916, continuing to work until 1969, driving saws for the production of brush handles, sieve rings, coffins etc. It looked its age, with a large flywheel, disc crank and 4-bar crosshead guide. It had a jet condenser below, and a vertical air pump; this was driven from an overhung crank on the end of the main engine shaft, outside the flywheel and belt pulley, with a pitman beam to a bell crank which also operated the feed pump. The Meyer valve, it was said, was set to cut off at half stroke, regardless of the load. It was stopped very shortly after being photographed and was then moved to the Bressingham Steam Museum in Norfolk.

74. Hemel Hempstead, British Paper Co. Ltd., Frogmore Mills, Apsley End (TL058055)

Photo taken/Reference: 28.4.1988/569-37
Type: Inverted vertical twin cylinder (enclosed)
Maker and Date: Ashworth & Parker Ltd., Bury. No. 2127. 1956
Cylinder dimensions: 11ins x 7ins. Piston valves
Hp: c.130 (max) **Rpm:** 50–300 **Psi:** 100
Flywheel diameter: 4ft 6ins
Service: Driving dry end of paper machine via gearing and belts. Paper machine by J. Milne & Son, Edinburgh, 1895 (installed second-hand, 1907)

This was probably the last paper mill in the country with a working steam engine, still operating commercially until May 2000. When seen in 1988, the engine was driving the dry end of the paper machine, working 24 hours a day, 5 days a week. As was quite usual with paper mill drives, speed was controlled and varied by driving the Pickering governor via a pair of cone pulleys (as shown more clearly in *Stationary Steam Engines of Great Britain*, Volume 7, No. 52). The 50–300 rpm variation corresponded with a throughput of 30–200 ft/minute, with a speed nearer the upper end of this range probably more usual. The drive was by reduction gearing to a countershaft, from which belts drove in opposite directions along the length of the machine, with the drive continuing via further pulleys and belts. Exhaust steam at c.10–12psi went to the drying cylinders. Following its purchase by Dacorum Borough Council in 2002, the mill was leased to Paper Trail Ltd., the intention being to create a working museum.

71

72

73

75. Royston, T.W. Streader & Son Ltd., coffin manufacturers (TL347413)

Photo taken/Reference: 10.5.1981/266-13
Type: Compound overtype
Maker and Date: Richard Garrett & Sons, Leiston. Installed (second or third-hand) c.1955–6
Cylinder dimensions: 8¾ins/15¼ins x 1ft 5ins. Piston valves
Flywheel diameter: 7ft 6ins
Service: Driving alternator by belt from flywheel

This was Garrett's locomobile design, largely the work of one Emerich Smach, who had previously been employed by Heinrich Lanz of Mannheim, an earlier builder of this type (see R.A. Whitehead, *Garretts of Leiston*, 1964). It had piston valves, with the HP eccentric linked to a crankshaft governor. It appears to have been without a condenser and so would have been of the CSV (compound superheated) type rather than the CCS. It may have been last in use as a boiler only (the HP crank was wedged) and the superheater removed. Unfortunately, it has so far proved impossible to trace the early history of this engine. It is believed to have been purchased second-hand by Streaders, and may have been first installed by them at their Hatfield works (Streaders' works was moved from Hatfield to Royston c.1955–6). Following closure in 1965, the engine became increasingly derelict until all was swept away, probably later in 1981.

76. Wheathampstead, Benjamin Cole & Son, Hyde Mills (TL133170)

Photo taken/Reference: 16.6.1996/801-11
Type: Horizontal single cylinder
Maker and Date: Nathan Varty, Royston Iron Works. Installed c.late 1880s
Cylinder dimensions: c.12ins x 2ft. Slide valve, plus expansion slide valve
Rpm: c.63 (max)
Flywheel diameter: 9ft 6ins
Service: Driving (as alternative to water wheel) mill stones and/or roller mills – by belt from 5ft diameter pulley

There are two engines here, the arrangement of the millwork suggesting that this was probably the first, put in in the 1880s as a supplementary drive to the stones, but also capable of providing power for a now removed roller plant installed in 1893. (The slightly smaller, second engine, also by Varty, was probably installed, primarily to drive the roller mill, a little later.) A perhaps typical product of a small country foundry, there were interesting design features: the flat bed linked by tie rods to the air pump/jet condenser bed behind; the Laird or box type crosshead guide; oval-rimmed flywheel; and a variable expansion valve controlled by a hand-adjustable link. The horizontal spindle into the upper part of the valve chest was presumably for a throttle valve linked to a (Watt type?) governor which at some time was replaced by the existing Pickering type. The use of steam power probably stopped in the late 1930s, but the waterwheel remained in use until the mid 1990s. (For further information see "Steam Engines at Hyde Mills", *ISSES Bulletin*, Vol. 23, No.3, 2001.)

75

74

76

ISLE OF MAN

77. Douglas, Isle of Man Railway Company Workshops (SC375753)

Photo taken/Reference: 29.4.1985/478-29
Type: Horizontal single cylinder
Maker and Date: John Chadwick & Son, Prince's Bridge Ironworks, Manchester. Installed (second-hand) 1920
Cylinder dimensions: c.9ins x 18ins. Slide valve
Flywheel diameter: 5ft 6ins
Service: Belt drive from 3ft diameter pulley to overhead workshop lineshaft

This is an example of the sort of simple slide valve horizontal mill engine built in enormous numbers for small powers. It was bought in January 1920 from the well-known Bolton dealers, Thomas Mitchell & Sons for £80 to replace the previous workshop engine. With its large disc crank and 4-bar crosshead guide, it could well have been of late 19th-century date. Steam was provided from a locomotive boiler outside until, it is thought, around 1950, when it was superseded by an electric motor drive. It was removed in 1978 prior to installation in the Company's Port Erin museum, but then about five years later a commendable change of policy led to its being re-installed on its original bed.

KENT

78. Erith, Burt Bolton (Timber) Ltd., timber importers, Belvedere (TQ501804)

Photo taken/Reference: 10.3.1980/203-34
Type: Horizontal single cylinder vacuum pump
Maker and Date: Worthington Pump and Machinery Corporation, Laidlaw Works, Cincinnati, Ohio. No. L26753. 1935
Cylinder dimensions: 10ins x 12ins. Slide valve
Rpm: c.100 (max 300) **Psi:** 70–85
Flywheel diameter: 4ft (two flywheels)
Service: Standby vacuum pump for timber creosoting cylinder

There had been two engines side by side, acting as air/vacuum pumps for the creosoting process. The other had been replaced by an electrically driven pump by Reavell & Co., Ipswich. The remaining engine was used almost always as a standby vacuum pump, the air pressure required (75 psi) being usually impossible to obtain with the steam pressure available. The engine had a top speed of 300 rpm, at which point the "Erie" governor cut off the steam supply, but normally worked at about 100 rpm. As will be seen, the 11inch diameter pump was in tandem behind the steam cylinder and it was fitted with a "Laidlaw Feather Valve". The engine was dismantled in 1983, but there is no information to hand as to its subsequent fate.

79. Eythorne, Tilmanstone Colliery – No. 1 Pit (TR288505)

Photo taken/Reference: 20.11.1985/501-13
Type: Horizontal twin cylinder capstan
Maker and Date: Dunlop, Bell & Co. Ltd., Liverpool
Cylinder dimensions: c.8ins x 12ins. Slide valves; Stephenson link reverse
Service: Two drums for shaft servicing

In 1985, just before its final closure in the following year, Tilmanstone Colliery was notable for possessing five capstans or winches: one at No. 3 Pit – electrified; one at No. 2 Pit – converted to compressed air; and three at No. 1 Pit – one on compressed air, the other two, one of which is illustrated, still on steam. This was Dunlop, Bell's standard design, and similar to that of a number of other makers. The two drums (c.2ft 6ins diameter) had worm wheels on their inner ends, with the wormshafts each driven by reduction gearing from a single pinion on the crankshaft running underneath, between the two drums. Despite its derelict appearance, it was apparently still in use.

77

78

80. Sevenoaks, Sevenoaks Brick Works Ltd., Greatness (TQ535573)

Photo taken/Reference: 8.10.1974/19-36
Type: Horizontal single cylinder
Maker and Date: Robey & Co. Ltd., Lincoln. No. 33242. 1914
Cylinder dimensions: 20ins x 3ft. Drop valves; slide exhaust valves
Hp: 150 **Rpm:** 75 **Psi:** 80
Flywheel diameter: 12ft
Service: Driving mixers, presses and elevator, producing sand-lime bricks etc.

The largest of Robey's E Class single cylinder engines seen (see also illustrations Nos. 1 and 153), this had been built for the Inchicore Brick Co. Ltd., in Eire, from whom it had been purchased by the Sevenoaks company in 1933. After a rebuild here in 1933–4, it continued to provide good service until 1978. A new piston had been fitted in 1952, and it required further attention in 1970, when a replacement piston rod was machined and fitted by the Brick Works. In 1973 it was working 23 hours a day, 5 days a week. The steam valves were controlled by Robey's patent drop valve gear, but the governor (unlike that of the Fairfield Hospital engine – see No. 1) was located alongside the cylinder. Drive was from the flywheel by a belt 16ins in width and 76ft 6ins long to a 5ft diameter second motion shaft pulley. The sand-lime bricks produced, white in colour, were made not from clay but from a mixture of damp sand and hydrated or slaked lime. This was compressed into bricks, which were hardened in an autoclave by being subjected to high-pressure steam. The single oil-fired Lancashire boiler was capable of supplying steam at 135 psi. The engine was removed for preservation in 1978 and is believed to be now at the Amberley Chalk Pits Museum, in pieces.

81. West Wickham, Metropolitan Water Board, West Wickham Pumping Station (TQ380646)

Photo taken/Reference: 29.9.1968/3228
Type: Inverted vertical triple expansion pumping engine
Maker and Date: Hathorn, Davey & Co. Ltd., Leeds. No. 7261. 1923
Cylinder dimensions: 16ins/28ins/46ins x 3ft 6ins. Corliss valves
Hp: 205 whp **Rpm:** 25 **Psi:** 135
Flywheel diameter: 13ft
Service: Water supply to Farnborough Reservoir

West Wickham was one of the works which the Metropolitan Water Board inherited from the Kent Water Works Company in 1904. The then existing engine by John Penn (see *Stationary Steam Engines of Great Britain*, Volume 8, No. 48) was in 1923 augmented by this Hathorn, Davey engine. The Corliss valves were operated by Craig's valve gear, with a modified Proell-type governor to control the HP trip gear. The two well pumps were single-acting bucket and plungers (17ins x 3ft 6ins) operated by quadrants driven by a connecting rod from an extension of the engine crankshaft. Three high level ram pumps (13½ins x 3ft 6ins) were driven from the engine crossheads. Originally pumping about 2.25 m.g.d. with a total head of 350ft, the yield later declined to about 1.5 m.g.d. The two Lancashire boilers of 1922 and 1937 were by Ruston & Hornsby Ltd. and Edwin Danks, and were hand-fired. The engine stopped work on 15 March 1969, and all buildings and machinery were demolished.

80

81

LANCASHIRE

82. Backbarrow, Charcoal Iron Company Ltd., Backbarrow Iron Works (SD355847)

Photo taken/Reference: 26.4.1974/13-18
Type: Horizontal single cylinder blowing engine
Maker and Date: Installed (second-hand) c.1924
Cylinder dimensions: c.18ins x 3ft. Slide valve; separate expansion valve
Flywheel diameter: 12ft 6ins
Service: Blast furnace blowing engine. Blowing cylinder (c.36ins dia) in tandem behind steam cylinder

The furnace, built in 1711, continued in use as the last in Great Britain using charcoal until 1920, when coke-firing was introduced by its final owners. In its last years, until 1966, it specialised in the production of cold blast pearlitic iron, used by many of the leading roll makers. Little has been discovered about the history of its engine or its working conditions. The Pickering governor, which disappeared about 1980, was made by Tangyes, but it seems unlikely that this company also made the engine. It probably worked condensing – in an adjacent building are the remains of what is presumed to have been a jet condenser, probably working with an independent air pump. In its later years, the engine was apparently used as an alternative to gas engine and electrically-driven blowers, and perhaps completely out of use by the late 1950s. Unfortunately, this important site and its engine have become increasingly ruinous over the years.

83. Bold, Sutton Manor Colliery (SJ518908)

Photo taken/Reference: 11.11.1979/194-25
Type: Horizontal cross compound compressor
Maker and Date: Walker Brothers Ltd., Pagefield Ironworks, Wigan. No. 17580. 1912
Cylinder dimensions: 25ins/41ins x 4ft. Corliss valves
Hp: 345? **Rpm:** c.30–65 **Psi:** 150
Flywheel diameter: 16ft 6ins
Service: Providing 5,000 cubic ft/minute of compressed air at 80 psi

Five steam engines could be seen at work here until the 1980s, the last being the Yates & Thom winder (see *Stationary Steam Engines of Great Britain*, Volume 3.2, No.122) which only stopped in 1991, when the Colliery closed. This Walker Brothers compressor was one of two identical engines built for Richard Evans & Co. Ltd. in 1912, the other being for that Company's Lyme Pits. Corliss valves, with separate eccentrics for inlets and exhausts were fitted, the inlets with Dobson trip gear, the HP being controlled by a Porter governor. Two stage air cylinders (26inch and 41inch) were in tandem behind the steam cylinders. There was another, larger, Walker Brothers compressor dating from 1939–43, which unlike the 1912 engine could be worked condensing. This was wrecked on 24 October 1980, from which date the 1912 compressor, previously standby, was until 1985 in continuous use. In 1992 it was removed to Chatterley–Whitfield Colliery for preservation.

84. Bold, Sutton Manor Colliery (SJ518908)

Photo taken/Reference: 17.10.1982/364-36
Type: Horizontal cross compound (fan) engine
Maker and Date: Walker Brothers Ltd., Pagefield Ironworks, Wigan. No. 15744. 1910
Cylinder dimensions: 22ins/36ins x 3ft 6ins. Corliss valves
Rpm: 102 **Psi:** 150
Flywheel diameter: 16ft
Service: Mine ventilation. Driving by 14 ropes from flywheel Walker's "Indestructible" fan, 25ft diameter and 9ft wide. Capacity 500,000 cubic ft of free air/minute

Like the 1912 compressor, this was a non-condensing engine, with Dobson trip gear and a Porter governor. By the 1970s it was on standby only and working at a reduced speed of around 50 rpm (fan c.70 rpm) or less. However, it was notable for being the last steam fan engine to be used on a British colliery – in July 1985, while repairs were carried out to the main ventilating plant. In 1986, the engine and fan were removed to Trencherfield Mill, Wigan for preservation. There were also here two turbine-driven alternators: one a mixed pressure turbine (by Metropolitan Vickers, 2,000 kW) which could take the exhaust steam from all the reciprocating engines; the other a high-pressure turbine (Hick, Hargreaves, 3,125 kW) which proved to be something of a "white elephant" as a result of its heavy steam consumption. The boiler plant included eleven Lancashires, some burning mine gas, and, disused by 1983 and demolished soon afterwards, a German multi-storey water tube boiler of immense size, burning coal waste.

82

83

84

85. Bolton, Stephen Glaister Ltd., dyers & finishers, Red Bridge Mills, Ainsworth (SD753099)

Photo taken/Reference: 16.8.1977/81-19
Type: Diagonal twin cylinder
Maker and Date: D. Stewart & Co. (1902) Ltd., Glasgow
Cylinder dimensions: c.6ins x probably 10ins. Slide valves
Hp: 15? **Rpm:** c.40 (variable) **Psi:** 90
Flywheel diameter: 35ins
Service: Driving a set of drying cylinders

This was a typical example of the type, with slightly offset steam cylinders in an inverted vee arrangement, with a single crank and a single eccentric for both slide valves, and single slipper guides. Two columns of drying cylinders (or "tins") were driven by reduction gearing and then from the second motion shaft: (1) a second pair of reduction gears to one column, and (2) a chain drive to another pair of reduction gears to the other. As usual, the steam exhausted to the drying cylinders. The engine presumably remained in use until the company went into liquidation at the beginning of May 1985. It was then scrapped, although a second, similar spare was reportedly removed for preservation.

86. Bolton, George Taylor (Brass Founders) Ltd., All Saints Street Works (SD717097)

Photo taken/Reference: 27.10.1982/372-32
Type: Inverted vertical single cylinder
Maker and Date: Tangyes, Birmingham. No. 3706. c.1870s?
Cylinder dimensions: c.10ins x 12ins. Slide valve
Rpm: 140?
Flywheel diameter: 4ft 6ins
Service: Reportedly driving machines (lathes etc.) via two lineshafts

The company established by James Marsden around 1840 started the All Saints Street Works as a brass cock foundry in the early 1850s. Caywood & Taylor took over c.1880 or just before, and by c.1890 George Taylor (who had been Marsden's foreman) was on his own. The engine was thought to have been installed when the works started, which may not be too far from the truth – the Tangyes serial number suggests the 1870s as a likely building date. Quite massively built for such a small engine, the drive was from the flywheel, grooved for 4 ropes. The engine speed shown is a maximum calculated from the Pickering governor, which was probably a replacement of a Tangyes original – the throttle and stop valve casting which forms the governor base is inscribed: THE "SOHO" PATENT LAW & PINKNEY'S. The engine was non-condensing, but had an oil separator on the exhaust pipe. It probably stopped working in the 1950s, and by 1994 the works had become S. & S. Engineering Co.

85

86

87. Chadderton, Courtaulds Ltd., Manor Mill, Victoria Street (SD911058)

Photo taken/Reference: 6.7.1982/338-10
Type: Horizontal twin cylinder (fire) pump
Maker and Date: George Mills & Co., Radcliffe. Probably c.1906
Cylinder dimensions: 14ins x 15ins. Slide valves
Flywheel diameter: 4ft 4ins
Service: Mill fire pump

George Mills & Co. Ltd., of the Globe Iron Works, Radcliffe specialised in the manufacture of fire pumps, as well as sprinklers and other fire extinguishing appliances, and this was one of the more substantial examples surviving into recent years. Its assumed date of manufacture is that of the construction of the mill by Manor Mill Ltd. As usual the slide valve chests were within a single casting between the steam cylinders, and the double-acting pumps, 8ins in diameter, were in tandem behind. The flywheel was cased in with metal sheeting, which was unusual. The maker's plate described it as a "steam fire and boiler feed pump", and apparently the latter use had been possible in an emergency. It had been disused since around 1970, but was still when seen being kept in reasonable condition by the mill's owners – by then Courtaulds Ltd. Unfortunately, within about a year, it had been scrapped.

88. Chorley, British Leyland (U.K.) Ltd., Chorley Works, Grime Street (SD587166)

Photo taken/Reference: 12.11.1979/195-36
Type: Horizontal cross compound
Maker and Date: Clayton, Goodfellow & Co. Ltd., Blackburn. No. 566-7. 1914
Cylinder dimensions: c.18ins?/c.32ins? x 3ft 6ins. Corliss valves
Hp: c.600 **Rpm:** 68¾
Flywheel diameter: 16ft
Service: Driving by ropes from flywheel (10 grooves) to DC generator: Lancashire Dynamo & Motor Co. Ltd., Trafford Park. 460 kW. 220 rpm

This engine had a slightly unusual history, having been first installed in 1914 for driving a newly built cotton weaving shed, which was in fact never used as such, although the looms were installed. The shed was laid out with lineshafts – presumably electric motor driven – and so represented an intermediate stage between engine-driven lineshafts and individual motor drives. After the 1914–18 War, the works was purchased by Leyland Motors Ltd., who used the engine, probably until 1938. It then remained workable, being it is said steamed every Christmas until 1963 at the instigation of Sir Henry Spurrier (managing director 1949–63, latterly also chairman, of the company). With steam inlet valves controlled by Craig's gear, Whitehead governor and jet condenser/air pump behind the LP cylinder, it is very similar to the engine at Holmes Mill, Clitheroe (see *Stationary Steam Engines of Great Britain*, Volume 3.1, No. 99), although with a Corliss valve rather than slide valve LP cylinder. It was removed in 1985, and is now privately preserved in Derbyshire.

89. Darwen, Shaws of Darwen Ltd., tile manufacturers etc., Whitebirk Works, Waterside (SD712232)

Photo taken/Reference: 20.10.1982/373-12
Type: Inverted vertical compound (enclosed)
Maker and Date: James Howden & Co., Glasgow. No. 235. 1906
Cylinder dimensions: 11ins/18ins x 8ins. Piston valve
Hp: 190 bhp(max) **Rpm:** 450 **Psi:** 120
Flywheel diameter: 3ft 9ins
Service: Driving grinding pan and clay mixers

Established by Shaws Glazed Brick Company Ltd. in 1908, the Whitebirk Works was by the 1970s being operated by Shaws of Darwen Ltd., with two high-speed steam engines still in use, exhausting to the drying floors. These were a Browett, Lindley compound of 1950, driving a 60 kW DC generator, and the central valve Howden illustrated, standing (very suitably) on a base of glazed bricks. A record of repairs etc. included the fitting of the James Tate stop motion in 1939. It was still in 1977 driving a section of the works via a lineshaft driven by 6 vee belts from a 30 inch diameter pulley, out of the photograph to the right. Formerly it had also driven another 60 kW DC generator (removed by 1977) – possibly from the 45 inch pulley grooved for 7 belts visible in the photograph. Probably in July 1980, use of the engines had ceased, and they were removed in 1982 – possibly for eventual preservation.

87

88

89

90. Leigh, Bickershaw Colliery (SJ633997)

Photo taken/Reference: 7.9.1980/233-5
Type: Horizontal cross compound compressor
Maker and Date: Walker Brothers (Wigan) Ltd., Pagefield Ironworks, Wigan. No. 23603-4. c.1919
Cylinder dimensions: c.22ins/c.42ins x 4ft 6ins. Corliss valves
Rpm: 55 **Psi:** 110
Flywheel diameter: 16ft
Service: In intermittent use (in 1974) as a booster to maintain the air pressure at 70–80 psi, providing c.1500 cubic ft/min

Three Lancashire collieries continued to use steam plant until 1980, and the oldest of them was Bickershaw, the four shafts of which were sunk in 1872–81. Additional to the two winders (see *Stationary Steam Engines of Great Britain*, Volume 3.1, No. 146) was this compressor. Walker Brothers specialised in air compressors and ventilating machinery – particularly for collieries, and this was a typical example of the period – with Corliss valves, Dobson trip gear, and Porter governor. Between the steam and air cylinders were substantial distance pieces, inscribed with the maker's name. The original stated capacity of the two-stage air cylinders (26ins and 42ins in diameter) was 4,390 cubic ft/minute at 55 rpm, but when seen in 1974 it was working at around 16-20 rpm and looking neglected. It was disconnected around 1977, but lingered on until 1982/3 before being scrapped. The two winding engines were stopped and scrapped in 1980.

91. Liverpool, The Mersey Docks and Harbour Company, Huskisson Dock Impounding Station (SJ332930)

Photo taken/Reference: 26.4.1978/108-17
Type: Inverted vertical tandem compound engines
Maker and Date: J. & H. Gwynne & Co., Hammersmith. c.1900
Cylinder dimensions: 22½ins/42ins x 30ins. Slide valves
Psi: 100
Flywheel diameter: 9ft
Service: Each engine direct-coupled to Gwynne 51in "Invincible" centrifugal pump (i.e. with suction and delivery pipes 51 inches in diameter). Pumping water from River Mersey into the Dock

The impounding system involved maintaining the water level in the Dock at not lower than the level of the high water of a low spring tide, the pumps being designed to make good the water lost through lockage and leakage. The three pumps shown in the photograph, installed by the Mersey Docks and Harbour Board in around 1900, had a combined capacity of 1,200 tons/minute. Not shown is the fourth Gwynne set of only slightly different design, first installed at Coburg Dock in 1887, and transferred here in 1908. It had its slide valves in tandem, unlike those of the three engines illustrated; and its surface condenser was built into the back frame (those of the other three were below floor level). The engines may have been in use until c.1960, and survived until c.1987, when one engine (without its pump) was removed for preservation and the remainder were scrapped.

90

92. Liverpool, Ogden's (Imperial Tobacco Ltd.), Boundary Lane (SJ366917)

Photo taken/Reference: 3.8.1988/598-30
Type: Inverted vertical compound (enclosed)
Maker and Date: Browett, Lindley & Co. Ltd., Patricroft. Nos. 2994 and 2995. 1923
Cylinder dimensions: c.16ins/c.24ins x ?. Piston valves
Hp: 430 bhp **Rpm:** 375 **Psi:** 150
Flywheel diameter: 5ft
Service: Two engines, each direct-coupled to Mather & Platt Ltd. 300 kW alternator

Ogden's became part of Imperial Tobacco in 1902 and established this factory to manufacture pipe tobacco. The two Browett engines stood side-by-side and were presumably installed new. They were two valve compounds, with the HP piston valve at one end and the LP valve between the two cylinders. The HP valve, with Browett, Lindley's patent variable expansion gear, could be rotated by a linkage with the governor. There were three Ruston Paxman "Thermax Two" oil-fired boilers, which were 1967 replacements of the original five Lancashires. The engines had been last brought into serious and continuous use during the 1974 three-day week period, but although nominally still standby as late as 1993, they could only accommodate a very small part of the factory load, and were probably little used.

93. Manchester, Joseph Holt Ltd., Derby Brewery, Cheetham (SJ841996)

Photo taken/Reference: 5.7.1982/336-6
Type: Horizontal single cylinder
Maker and Date: Robey & Co. Ltd., Lincoln. No. 50390. 1946
Cylinder dimensions: 11ins x 15ins. Piston valve
Hp: 25 **Rpm:** 140 (des. max) **Psi:** 100
Flywheel diameter: 5ft
Service: Driving sack hoist. Originally drove all brewery equipment

Robeys were one of the few British companies still building reciprocating steam engines into the 1950s, and this is a relatively late example, delivered here new on 17 May 1946. It was quite compact, with a bedplate and trunk guide in a single casting, with overhung, cased-in, cylinder and piston valve bolted on to the rear end. There are two bearings on the bed with double web crank between, flywheel at one end of the crankshaft and belt pulley, 3ft diameter, at the other. Unfortunately, from 1982 its only use was as a standby drive to the sack hoist, and by 1989 it was completely disused (and indeed covered in rubbish).

94. Manchester, United Ship-Repairers & Engineers (Manchester) Ltd., Manchester Dry Docks, Trafford Wharf Road (SJ798974)

Photo taken/Reference: 5.7.1982/337-6
Type: Inverted vertical single cylinder engines (two)
Maker and Date: Tangyes Ltd., Cornwall Works, Birmingham. 1893
Cylinder dimensions: 20ins x 18ins. Slide valve; Meyer expansion valve
Hp: 120 ihp (max) **Rpm:** 100–120 **Psi:** c.100
Flywheel diameter: 5ft
Service: Each driving centrifugal pump, 36ins diameter, suction lift 10ft, for emptying Nos. 1 & 2 Dry Docks. (There was no discharge head – the water level on the outlet side was below the pump discharge.)

The engines and pumps were supplied by Tangyes for the Manchester Ship Canal's dry docks in 1893, and remained in use for the next hundred years. They are a handed pair, arranged end-on, with direct-coupled pumps at opposite ends, strongly built with double web counterbalanced cranks and single slipper guides on the back supports. Steam came from two oil-fired Economic type boilers and the engines exhausted to atmosphere via a feedwater heater tank. With both pumps running, a full dock (Nos. 1 and 2 were 500ft & 525ft x 65ft x 25ft deep) could be emptied in 2–2$^{1}/_{2}$ hours. (The use of the slightly larger No. 3 Dock had ended c.1980.) The engines continued to work as required at least until 2002 when, after one or more changes in ownership, the docks appear to have been closed. The current position regarding the engines is unknown.

91

92

93

95. Manchester, Water Street Hydraulic Power Pumping Station (Manchester Corporation Waterworks) (SJ833983)

Photo taken/Reference: 30.6.1972/4149
Type: Inverted vertical triple expansion pumping
Maker and Date: Galloways Ltd., Manchester. 1909
Cylinder dimensions: Cylinders (15ins/22ins/36ins x 2ft) removed 1928
Hp: (210 ihp) **Rpm:** (60) **Psi:** (135)
Flywheel diameter: Flywheel removed
Service: One of six engines, which were converted to electric motor drive, providing Manchester with hydraulic power. Each engine had three plunger pumps ($4^1/2$ ins in diameter) driven from the crossheads. Two accumulators (18ins diameter rams) provided water to consumers at 1,100 psi

This was the Corporation's third hydraulic power station, erected in 1909. The engines appear to have been of the Edward Ellington design associated with the Hydraulic Engineering Co. Ltd., of Chester, who presumably contracted out their actual building to Galloways. In 1928 they were converted to electric motor drive, steam cylinders, valve gear and flywheels being removed, and 220 hp DC motors connected to the crankshafts by reduction gearing. As they formed part of the frame, the surface condensers remained, and one of these can be seen in the illustration. The station was closed on 28 December 1972, thus bringing Manchester's hydraulic power system to an end. Five of the engines were scrapped, but the other is now in the Manchester Museum of Science and Industry.

96. Oldham, Raygrade Ltd., Royd Mill, Chamber Road, Hollinwood (SD913034)

Photo taken/Reference: 20.10.1982/372-19
Type: Horizontal twin cylinder rotative (fire) pump
Maker and Date: Joseph & James Hall, Oldham. 1907?
Cylinder dimensions: c.8ins x 12ins. Slide valves
Flywheel diameter: 4ft
Service: Mill fire pump. Double-acting pumps, 6ins diameter, in tandem behind steam cylinders

This large cotton spinning mill was erected in 1907 and the fire pump may have been installed then by Alfred and Arnold Murgatroyd who formed Royd Mill Ltd. Joseph & James Hall were established in 1848 according to the small plate on the top cover of the slide valve chest, a single casting between the two cylinders. Whether they actually built the engine is perhaps not absolutely certain. They describe themselves as "fire engineers" and among their other related products were fire escape staircases. The mill was taken over by the Lancashire Cotton Corporation in the 1930s, and later by Courtaulds who ceased production here in 1981. Somewhat derelict when photographed, the pump was probably removed around 1988-93. It is now at the Astley Green Colliery preservation site.

97. Radcliffe, (Ex) Whittakers of Mount Sion Ltd., Mount Sion Bleachworks (SD768068)

Photo taken/Reference: 16.8.1977/81-26
Type: Stationary wharf crane
Maker and Date: Thomas Smith, Rodley, Leeds. No. 3184. Probably early 1880s
Cylinder dimensions: 7ins x11ins (vertical twin). Slide valves with reversing links
Service: Transfer of coal from canal boats into the Mount Sion Works

The crane, on the bank of the Manchester, Bolton & Bury Canal, and the Mount Sion Works occupy a site between the Canal and the River Irwell. Taking over an existing calico printing works, John Whittaker & Co.'s bleachworks was established in 1859, and it was presumably Whittakers who installed the crane, possibly in the early 1880s – this is the building date suggested by an order number of 3184, the cast-on number. The crane may have become disused in the 1930s; by 1977 the works was occupied by Alphacell Ltd., processors of manilla hemp for paper-making, and the crane was completely derelict. As will be seen, hoisting was via two pairs of reduction gears to the winding barrel. Slewing was via a second motion shaft carrying two facing bevels providing alternative drives to a vertical shaft which led via further double reduction to a pinion engaging with the circular inward facing slewing rack. The vertical boiler was by Marshalls of Gainsborough.

94

95

96

98. Warrington, Greenall Whitley & Co. Ltd., Wilderspool Brewery, Greenalls Avenue (SJ612866)

Photo taken/Reference: 25.4.1978/105-32
Type: Horizontal single cylinder (refrigerating) engine
Maker and Date: Davey, Paxman & Co. Ltd., Colchester. No. 11179. 1902
Cylinder dimensions: 19ins x 2ft. Slide valve; variable expansion slide valve
Hp: 55 ihp* **Rpm:** 75 **Psi:** 60
Flywheel diameter: 8ft 6ins
Service: Ammonia compressor for refrigeration process (At design steam pressure of 45 psi and 65 rpm.)

The Linde mechanical refrigeration process involves the compression and liquefying of ammonia gas which is then allowed to evaporate. This was a rare (possibly unique) surviving example of one of the steam engine driven ammonia compressors which from 1890 Davey, Paxman supplied to The Linde British Refrigeration Co. Ltd. in large numbers. It consisted of a girder frame engine cross-coupled to the compressor, which had a cylinder 10ins diameter by 21ins stroke. It had what was described as a "Davey, Paxman new adjustable high speed type" governor, which controlled the variable link connected to the expansion valve. Davey, Paxman was still supplying it with parts as late as 1947, but it may soon after this have become disused. The spacious engine room also contained a pair of single cylinder Thornewill & Warham engines for driving the brewery machinery. These stopped working in 1962, but all was kept in immaculate condition until, very regrettably, all was dispersed after the brewery closed in 1991.

99. Wigan, Jeffreys, Miller & Co. Ltd., malt extract manufacturers, Leyland Mills (SD585076)

Photo taken/Reference: 9.8.1983/426-36
Type: Horizontal single cylinder pump
Maker and Date: Manlove, Alliott & Co. Ltd., Nottingham. 1921
Cylinder dimensions: c.10ins x 20ins. Slide valve
Rpm: c.60–70 **Psi:** Variable
Flywheel diameter: 5ft
Service: Pump for creating vacuum for two evaporators used to concentrate malt extracts used for home brew kits and bakeries

The steam from the two evaporators, also by Manlove, Alliott, was condensed in a column, through which the pump pulled water drawn from the River Douglas. This water together with the condensate was then ejected and returned to the river. The evaporating capacity was 500 gallons/hour and the pump was capable of pumping 45 gallons/minute including the condensate. The pump (c.12ins in diameter) is at the opposite end to the steam cylinder and was driven by side rods passing below the crankshaft to each side of the double web crank. The engine probably once had a Pickering governor, by then removed. The exhaust steam could, if required, be used to heat the evaporators. The flywheel was within the frame, but there was an overhung belt pulley which in the past had apparently been used to drive some of the machinery via the lineshaft above. Use of the engine was discontinued around the end of 1998.

98

97

99

LEICESTERSHIRE

100. Broughton Astley, Broughton Astley Tileries Ltd., roofing tile manufacturers (SP535919)

Photo taken/Reference: 22.6.1974/16-9
Type: Horizontal single cylinder
Maker and Date: Ruston, Proctor & Co. Ltd., Lincoln. No. 15736. 1891
Cylinder dimensions: 16ins x 32ins. Slide valve; Meyer expansion valve
Flywheel diameter: 10ft
Service: Drove brickworks machinery

This was a Ruston, Proctor Class "B" engine, originally sold on 27th May 1891 to the Gas Light & Coke Company. By 1973 it lay derelict amid the ruinous remains of the works, in an engine room as often as not up to a foot deep in water, as seen here. So far it has not been possible to discover anything about the history of the works, and there seems to be no record of the engine's working conditions. (The maker's quoted rating for this type and size was evidently 65 hp with 80 psi steam and 70 rpm.) It was a heavily built engine, with an overhung disc crank within the frame, 4-bar crosshead guide, and a Porter governor to control the throttle valve. The driven machinery had gone, although still remaining was the lineshafting, driven by an 18 inch leather belt from a 4ft diameter pulley next to the flywheel. There were also a few remnants of the two boilers (Lancashires?). By 1996 the site, which had been occupied by a scrap metal business, was being developed for housing, and it is believed that the engine was removed for private preservation.

101. Leicester, William Gimson & Sons Ltd., timber importers, Swan Lake Mill (SK578033)

Photo taken/Reference: 5.4.1995/763-16
Type: Inverted vertical compound (enclosed)
Maker and Date: Belliss & Morcom Ltd., Birmingham. No. 11157. 1953
Cylinder dimensions: 10ins/16ins x 7ins. Piston valves (V-type)
Hp: 156 bhp (des.) **Rpm:** 500 (des.) **Psi:** 150 (des.)
Flywheel diameter: 3ft 3ins
Service: Drove 100 kW alternator

This was one of two Belliss & Morcom compounds driving alternators by vee belts, both installed here second-hand – No. 9627 of 1940, which stopped working c.1985; and No. 11157 of 1953, shown here, which stopped c.1980. It was slightly unusual in being operated with a surface condenser, for which the plumbing had been removed, but for which the drive to centrifugal pumps for circulating water and condensate extraction can be seen in the foreground. The HP piston valve was fitted with governor-controlled expansion gear. The alternator (removed by 1991) had been driven by 14 vee belts from a grooved pulley, and was presumably the alternator which had been similarly driven by the engine at its original location – timber importers W. Scott & Sons Ltd., also of Leicester. (No. 9627 had been first installed at the Swastika Laundry Ltd., Ballsbridge, Dublin.)

100

101

LINCOLNSHIRE

102. Grantham, Fenland Laundries Ltd., Belton Lane (SK918367)

Photo taken/Reference: 22.6.1974/16-3
Type: Horizontal single cylinder
Maker and Date: Marshall Sons & Co. Ltd., Gainsborough. No. 58294. 1912
Cylinder dimensions: 11ins x 2ft. Slide valve
Hp: 59 bhp **Psi:** 80
Flywheel diameter: 8ft 6ins
Service: Last used to drive generator

The laundry dated from 1878, but it has yet to be discovered whether or not the engine was installed here new in 1912. Of Marshall's Class KE with a Pickering type governor (Class K was the same but with Hartnell's automatic expansion gear), it had a girder/trunk frame and a separate stone bed for the cylinder. Drive would have been from the 5ft diameter belt pulley but it had apparently been out of use since 1946. In 1974 the engine room floor was about 4 inches deep in water as a result of flooding from the nearby River Witham, but it was preserved on site (and at times run under steam) until 1989, when the laundry was demolished, and the engine removed for preservation.

103. Sutton Bridge, Cross Keys Swing Bridge Pump House (TF480209)

Photo taken/Reference: 30.8.1982/352-7
Type: Horizontal twin cylinder rotative pumping engine
Maker and Date: Sir W.G. Armstrong, Whitworth & Co. Ltd., Newcastle upon Tyne. No. 720. 1894–7
Cylinder dimensions: (8$^1/_2$ins x 9ins. Slide valves; Meyer expansion valves) Removed
Rpm: (75) 49.43 – after electrification
Flywheel diameter: (4ft) Flywheel removed
Service: One of two engines providing hydraulic power to bridge. Each one had two double-acting force pumps (rams 1.59in dia, pistons 2.25ins dia, stroke 9ins). Capacity of both sets of pumps c.24 gallons/minute

Two small twin cylinder engines, Sir W.G. Armstrong, Whitworth Nos. 719 & 720, were installed here by the Midland & Great Northern Joint Railway to provide hydraulic power for the combined road and rail swing bridge over the River Nene. The illustration shows one of the pumps, after their 1936 electrification, which involved the removal of steam cylinders, flywheel etc., and addition of an induction motor geared to the crankshaft through a David Brown reduction box. In this form, they continued to work until April 1988, pumping to two accumulators, providing water at 750 psi to the bridge. This had two hydraulic motors (each with three oscillating cylinders) located below the control cabin over the centre of the swing span; and four hydraulic jacks for lifting the bridge before it could be swung. The bridge is still hydraulically operated, with a motor-driven pump slung under the bridge deck. The pump house, with its disused pumps and accumulator tower, remains.

LONDON

104. Bethnal Green, Bethnal Green Baths, Cheshire Street (TQ343823)

Photo taken/Reference: 22.7.1978/119-10
Type: Horizontal single cylinder
Maker and Date: Marshall Sons & Co. Ltd., Gainsborough. Installed 1899?
Cylinder dimensions: 10ins x 14ins. Slide valve
Flywheel diameter: 5ft
Service: Drove laundry machinery, presumably via lineshaft by belt from 2ft 8ins diameter pulley

The Bethnal Green Baths, a splendid building visible from the railway just north of Liverpool Street Station, was built in 1899 and, sadly, closed in 1978. There were two small horizontal engines which had stopped work, probably in the 1960s – a Tangyes for driving an artesian well pump, and the Marshall, said to have driven two washing machines and two hydro-extractors (by then removed). The Marshall is presumed to have been installed when the Baths were built (an engine in this position is shown in the architect's plan). It appears to be of that company's "A" Class, with both bearings on the frame and an overhung cylinder, stroke about 1$^1/_2$ times the bore. Both engines were removed for preservation in 1985, and are now in the care of the Museum in Docklands.

102

103

104

105. Bow, North Thames Gas Board, Bow Common Gasworks (TQ371820)

Photo taken/Reference: 20.12.1973/7-3

Type: Horizontal twin cylinder rotative pumping engine

Maker and Date: Tannett, Walker & Co., Leeds. Possibly c.1880s.

Cylinder dimensions: 14 3/8 ins x 20ins. Slide valves

Flywheel diameter: 6ft

Service: Pumping to accumulator for provision of hydraulic power to cranes in the coal yard

A number of engines, out of use since around 1968, remained here into the 1970s, some as late as 1976 when the works was demolished. These included two pumping engines for providing hydraulic power, presumably installed by the Gas Light & Coke Company which had taken over in 1870. The Tannett, Walker was the earliest, put in for the two oldest cranes (which had been scrapped by December 1973); the other, by Sir William Arrol & Co. Ltd., Glasgow, was put in to provide additional power for a larger crane built by the East Ferry Road Engineering Works, erected around the late 1920s. The Tannett, Walker was an interesting design, with the pumps in tandem in front of the cross coupled steam cylinders. Each of these had double piston rods with forward extensions to the pumps – two on each side; and from each crosshead there was a rectangular section connecting rod to a double web crank. The slide valve eccentric rods were driven from the outer ends of the crankshaft. Very derelict when photographed, it was removed to Leeds Industrial Museum a month or two later.

106. Stoke Newington, D. Politi & Sons Ltd., manufacturing confectioners (TQ335870)

Photo taken/Reference: 27.2.1978/101-6

Type: Horizontal single cylinder

Maker and Date: Marshall Sons & Co. Ltd., Gainsborough. No. 36080. 1901

Cylinder dimensions: 5 1/2ins x 8ins. Slide valve

Psi: c.60

Flywheel diameter: 3ft

Service: Belt drive via lineshaft to stirrers in boiling pans

Politi's factory was established at 10 Manor Road in 1911, to manufacture confectionery and (notably) Turkish Delight. There were two small engines used for driving the stirrers in sixteen boiling pans – the Marshall and a Sisson single cylinder enclosed engine (No. 5370 of 1950), both installed second-hand – the latter used as a standby. (The Sisson had originally been used by Launceston & District Laundry, in Cornwall.) The Marshall was said to have been installed about 30 years previously (i.e. around the later 1940s) but so far nothing has been discovered about its previous history. It is very similar in design to the engine at Cheshire Street Baths – although smaller and with a Pickering type governor. Drive to the lineshafting was taken from a belt pulley 18ins in diameter. Until 1977 at least it had been worked about 8 hours a day in the period June–November and also at other times intermittently, but soon after this its use seems to have been discontinued. In August 1985, the factory was closed, the company having been taken over by Maynard's.

105

106

107. Westminster, Houses of Parliament (TQ303795)

Photo taken/Reference: 2.12.1989/643-26
Type: Horizontal twin cylinder rotative air compressor
Maker and Date: Marshall Sons & Co. Ltd., Gainsborough. c.1890
Cylinder dimensions: 9¼? ins x 14ins. Slide valves
Hp: 16 (nominal) **Rpm:** c.10-50 (variable) **Psi:** 80
Flywheel diameter: 5ft 6ins (two flywheels)
Service: Supplying air at 30 psi for the Shone hydro-pneumatic sewage ejector system installed by Hughes & Lancaster of Ruabon

When the Shone sewage system was installed in 1886, it was provided with air compressors driven by four of Atkinson's 4 hp patent differential gas engines. These were described as "somewhat noisy" (see "The Drainage System of the Houses of Parliament", *The Engineer*, Vol. LXXIX, 21 June 1895), and after a period of co-existence were superseded by the combined steam engine and compressor. This consisted of a handed pair of Marshall "A" Class engines, 8 nhp each, direct-coupled to a twin disc crank driving the compressor (c.14ins x 20ins stroke) between them. The compressor section may have been built by James Simpson & Co., who apparently supplied the whole and presumably obliterated the Marshall details cast on the slide valve covers. Two of the original three Shone ejectors are still in place and in use, as also for standby is the steam engine.

108. Westminster, The London Hydraulic Power Company, Grosvenor Road Pumping Station (TQ298781)

Photo taken/Reference: 16.5.1973/2-16
Type: Inverted vertical triple expansion pumping engine
Maker and Date: The Hydraulic Engineering Co. Ltd., Chester. 1910
Cylinder dimensions: 15ins/24ins/39ins x 27ins. Piston valve (HP); slide valves (IP & LP)
Hp: (312 ihp originally) **Rpm:** 49.8 (63 originally) **Psi:** (180)
Flywheel diameter: Flywheel removed
Service: Three single-acting ram pumps, pumping 360 gallons/minute to accumulator at 850 psi. (This was one of four engines remaining of the original six which were converted to electric drives.)

Grosvenor Road was the last pumping station to be built by the London Hydraulic Power Company. Around 1953 its six steam engines were converted to electric drives, with flywheels and valve gear being removed, and AC motors, geared 10:1 to the crankshaft, added. The cylinders and piston rods were retained to act as guides. In 1971 Grosvenor Road closed (the engines were scrapped around June 1973), and Wapping was left to carry the remaining load. On 30 June 1977 all pumping operations by the Company came to an end.

109. Westminster, The Strand Palace Hotel (TQ305808)

Photo taken/Reference: 14.3.1979/145-10
Type: Inverted vertical triple expansion (enclosed)
Maker and Date: Belliss & Morcom Ltd., Birmingham. No.7553. 1926
Cylinder dimensions: 12½ins/17ins/24ins x 12ins. Piston valves (HP with governor-controlled expansion gear)
Hp: 440 bhp **Rpm:** 350 **Psi:** 200
Flywheel diameter: 5ft 10ins
Service: Electricity generation

In 1926, Belliss & Morcom supplied three steam engine driven generator sets to J. Lyons & Co. Ltd. for The Strand Palace Hotel, then under construction: Nos. 7554 and 7555 were of their B3 (three cylinder double-expansion) type, of 330 bhp – presumably removed or scrapped many years ago; No. 7553, Belliss & Morcom Type T5, remained in full-time use probably until c.1965 and then on standby until c.1980 (it had been used in 1974 during the miners' strike). The two original Yarrow boilers – oil-fired, three drum, small tube type – also remained. They were originally designed for a working pressure of 250 psi with superheat to 550°F. (A suggestion that all three engines and the boilers were originally destined for a Royal Navy cruiser under construction at Wallsend Slipway has not been confirmed!) No. 7553, which exhausted at 20 psi back pressure to the hotel water heating system, was direct-coupled to a Crompton & Co. Ltd. 300 kW DC generator. It was still on site in the year 2000, below ground level in what was the Hotel's works department at 20 Tavistock Street, and as far as is known has survived the subsequent construction of new buildings above.

107

108

109

NORFOLK

110. Holkham, Holkham Estate workshops and sawmill, Longlands Farm (TF880400)

Photo taken/Reference: 26.7.1974/17-5

Type: Single cylinder overtype (or semi-portable)

Maker and Date: Richard Garrett & Sons Ltd., Leiston. No. 32587. 1914

Cylinder dimensions: 7ins? x 12ins. Piston valve

Rpm: 200 (des.)? **Psi:** 190 (des.)

Flywheel diameter: 4ft 2¾ins

Service: Driving sawmill and workshop machinery

The Holkham Estate provides an outstanding example of Victorian industrial farming, with its workshops at Longlands Farm dating from 1859–60. The existing engine was ordered for Lord Leicester in June 1914 and is described in the maker's records as Garrett's "SSWIV single cylinder... non-condensing semi-stationary engine" – with uptake superheater, Wilkinson waterheater and extended firegrate (the latter appears to have been removed). As would have been usual with this type, the piston valve cut-off was controlled by a Turner-Hartnell Patent crankshaft governor; the exhaust steam was allowed to escape into the chimney stack. The machinery was driven from a belt pulley (30¼ins diameter) via a lineshaft extending along the considerable length of the sawmill and workshops. The engine ceased work as a result of the sawmill being stopped – probably in 1962 (the last inspection of the multi-tubular boiler appears to have taken place on 27th November 1961).

111. Marham, Wisbech & District Water Board, Marham Pumping Station (TF717107)

Photo taken/Reference: 26.7.1974/17-18

Type: Horizontal duplex tandem triple-expansion non-rotative pumping

Maker and Date: Worthington-Simpson Ltd., Newark. No. 3108. 1938

Cylinder dimensions: 12ins/19ins/30ins x 2ft. Semi-rotary slide valves; HP cylinders each with two auxiliary cut-off valves

Hp: 205 ihp (des. max) **Spm:** 17–22 **Psi:** 175–200

Service: One of a pair of engines drawing water from shallow boreholes and the River Nar, pumping to a reservoir near Downham Market and supplying the Wisbech area. Two ram pumps (each engine) 15½ ins diameter x 2ft stroke, actually pumping 1.6 m.g.d. at 19 spm

Marham Pumping Station dated from 1903–8 and was reconstructed in 1938 when the two Worthington-Simpson engines (Nos. 3107–8) were put in. (Two Dodman side-by-side compounds of 1908 remained on standby until c.1962.) The Worthington triple-expansion direct-acting type was built in quite large numbers by James Simpson and their successors, Worthington-Simpson from 1889 – these were almost the last, being followed only by the two for Clacton Water Works in 1946–8, and they were the last to remain in use in a British waterworks. They were of the standard type in which the movement of the piston/pump rod on one side operated the main valves on the other via rocking shafts, one of which also had levers for driving the twin vertical air pumps for the surface condenser. The HP cut-off valves, linked to the piston rod movement on the same side, were adjustable by hand, although here apparently the setting was never altered. Electric pumps were installed in May 1976 and the engines were scrapped in the following year.

112. Norwich, New Mills Sewage Works Air Compressing Station (TG226091)

Photo taken/Reference: 4.9.1982/354-29

Type: Horizontal cross compound air compressor (parts only)

Maker and Date: Hughes & Lancaster, London & Ruabon. 1898

Cylinder dimensions: Cylinders (9½ ins/14½ ins x 18ins) removed

Psi: 120 (originally)

Flywheel diameter: 6ft (two flywheels)

Service: One of two identical engines supplying compressed air to six Shone's ejector stations around the City of Norwich, raising sewage to the main outfall sewer. (Hughes & Lancaster also manufactured the Shone's ejectors.) The double-acting air cylinders (15ins x 18ins stroke) were in tandem in front of the steam cylinders

In their original form, each compressor could alternatively be driven by a Francis type water turbine, connected by bevel gears to the crankshaft (with eccentric rod disconnected and piston rod between steam and air cylinders uncoupled). Each turbine, operating two air cylinders, when using 35,000 gallons of water/minute was able to compress 650 cubic feet of free air/minute to a pressure of 18 psi. (For an account of the plant as installed, see "Norwich Sewerage Works", *The Engineer*, Vol. LXXXVI, 22 July 1898.) In 1907 and 1909 electric motors, with chain drive to the crankshaft, were installed as a third possible power source, but in 1932 steam cylinders and related valve gear were removed. Presumably the Babcock & Wilcox water tube boiler with two refuse destructor furnaces also went out of use. The horizontal compressors may have become disused in 1968, but there were also three electrically-driven (vertical) compressors of various makes, dating from 1913–34, and presumably these at least remained in use until the Station shut down completely in 1972. It is being preserved, and is now perhaps in a state somewhat improved from that of 1982 when it was photographed.

110

111

112

NORTHAMPTONSHIRE

113. Northampton, St. Andrew's Hospital (Northamptonshire Area Health Authority) (SP770605)

Photo taken/Reference: 18.2.1989/607-33
Type: Inverted vertical compound (enclosed)
Maker and Date: W. Sisson & Co. Ltd., Gloucester. No. 6306. 1966
Cylinder dimensions: 10ins/16ins x 6ins. Piston valves
Hp: 150 bhp **Rpm:** 600 **Psi:** 140 (des.)
Flywheel diameter: c.40ins
Service: Electricity generation. Direct-coupled to 100 kW alternator by Crompton Parkinson Ltd., Chelmsford

High-speed steam generator sets were still being supplied to hospitals in the 1960s and indeed into the 1970s, particularly by Belliss & Morcom. (Possibly the final two were for Bangor Hospital around 1975. As far as is known, the last to remain in use – at Prince Charles Hospital, Merthyr Tydfil – stopped working in 1986.) The St. Andrew's Hospital Sisson replaced three earlier smaller Sissons dating from the early 1930s. It was their standard DC (double crank compound) type, with the two piston valves placed to one side of the cylinders and driven by a single eccentric with a rocking lever to a vertical rod forking to the two valve spindles. As was usual with hospitals, the exhaust steam (here at 3–4 psi back pressure) went to heating and hot water. After a relatively short working life, it was standby only from the early 1980s to around 1990 and then, completely superseded, removed to Markham Grange Nurseries/ Steam Museum, near Doncaster in 1993. Here it was installed and became capable of power generation by 1995.

113

114. Passenham, Harcros Chemicals Ltd. – Deanox Division, manufacturers of iron and red lead oxides (SP763396)

Photo taken/Reference: 30.6.1990/652-4
Type: Inverted vertical compound (enclosed)
Maker and Date: Browett, Lindley & Co. Ltd., Patricroft. Nos. 3222 & 3221. 1927
Cylinder dimensions: 16ins/24ins x 12ins. Piston valves
Rpm: 375 **Psi:** 160
Flywheel diameter: 5ft 1in
Service: Two engines, each direct-coupled to alternator, 300kW. In all, four generator sets provided about a third of the plant's power needs

The Deanshanger Oxide Works was started here as a family firm in 1935, being taken over by Harrison & Crosfield in 1983 and becoming part of Harcros Chemicals in 1988. By then the main product was "synthetic" iron oxide, used as a pigment for colouring bricks, tiles etc. Four Browett, Lindley engines were bought second-hand in the 1960s, the others being No. 2570 of 1915 and 3566 of 1947. All were brought onto their final site in 1973 – in a single newly commissioned Power House. They were said to have given extremely reliable service, although it will be seen that No. 3222 was out of commission when photographed. All were of the two valve type, with HP variable expansion gear – Nos. 3221–2 with governor control. Exhaust at up to 6 psi was to the heating vats. Unfortunately, in June 1999 the works was closed and the site was cleared. The engines, which had stopped in December 1996, are assumed to have been scrapped.

NORTHUMBERLAND

115. Allendale, Holmes Linn Shaft (Blackett Level), Sinderhope (NY842523)

Photo taken/Reference: 18.8.1981/299-8
Type: Diagonal twin cylinder winch
Maker and Date: Clarke, Chapman & Gurney, Victoria Engine Works, Gateshead. c.1875–85
Cylinder dimensions: c.5ins x 10ins. Slide valves; Stephenson link reversing
Service: Winding mineral from 192ft deep shaft

The Blackett Level was started at Allendale Town by the Blackett-Beaumont Company under the direction of Thomas Sopwith in 1855, the intention being to explore and drain the mines at Allenheads, 7 miles distant. The level was driven simultaneously from the portal at Allendale Town and from three shafts, one of which was Holmes Linn, 2$^{1}/_{4}$ miles from the portal. The main drive was suspended in 1903, at 4.68 miles from the portal, but work on a small scale seems to have continued until 1912. It has not been possible to find much information on this site. The steam winch is understood to have replaced a hydraulic engine in about 1900, and if so it must have been second-hand, the builder being known as Clarke, Chapman & Gurney (the name cast on to the bed and disc cranks) for only a relatively brief period, c.1875–85 (previously, they were Clarke, Watson & Gurney; afterwards Clarke, Chapman). The winch is located within part of the masonry bed of the hydraulic engine. The drum has a wooden barrel about 21 inches diameter and is driven via single or, alternatively, double reduction gearing. Possibly it was used intermittently until the 1920s or 1930s, and it was removed for private preservation in 1983.

116. Newcastle upon Tyne, Thomas Ness Ltd., St. Anthony's Tar Plant (NZ290631)

Photo taken/Reference: 19.8.1981/299-24
Type: Horizontal twin cylinder rotative pump
Maker and Date: Joseph Evans & Sons Ltd., Wolverhampton. No.35098. c.1944
Cylinder dimensions: 6ins x 9ins. Slide valves
Rpm: 45 **Psi:** 80
Flywheel diameter: 3ft 6ins
Service: Pump No. 1. Pumping pitch. Pump cylinders, 5$^{1}/_{8}$ins bore with 4$^{1}/_{2}$ins dia liner

This was one of a number of tar distillation plants operated by Thomas Ness, by then a wholly owned subsidiary of NCB (Coal Products) Ltd. The pump was an example of the rotative banjo type built by Joseph Evans which could be used for pumping thick liquids, particularly suitable for use in chemical and gas works. Twin cylinder, cross-coupled, it had double-acting steam-jacketed pumps, and two eccentrics on each side, including one each for the pumps' piston valves. Also here were two similar but single cylinder pumps, Nos. 2 and 3. All had stopped working – No. 3, pumping tar on the bank of the River Tyne, in the early 1970s; Nos. 1 and 2 on 31 May 1981, when the works stopped production. By 1984, all had been demolished, and the site cleared.

114

115

116

NOTTINGHAMSHIRE

117. Harworth, Harworth Colliery (SK625913)

Photo taken/Reference: 10.11.1979/191-15
Type: Horizontal twin cylinder capstan
Maker and Date: Robey & Co. Ltd., Lincoln. No.42064. 1924
Cylinder dimensions: 8ins x 1ft. Slide valves; Stephenson link reversing
Flywheel diameter: Brake wheel 28ins diameter
Service: Said to have been used as a platform sinking engine. Two drums c.4ft diameter and 3ft 6ins wide

This capstan, disused at Harworth for a number of years, had been re-assembled off-site c.1978–9, and when seen was connected up to a steam supply, its destination being the Lound Hall Museum. (Its present location is unknown.) Without markings, it was almost certainly built by Robey, their No. 42064, delivered to Markham & Co. Ltd., Chesterfield on 25th October 1924. (An identical engine, Robey No. 42063, was at New Monckton Colliery, Royston, and No. 42064 was the only other such engine found in Robey's records.) Probably it had been at Harworth from new – 1924 was the date the colliery was brought into production. As will be seen, the steam cylinders were side-by-side, valve chests between. The two drums, each with a 6ft 6ins diameter brake wheel, normally turn together and were driven by two wormshafts, each driven from the crankshaft by 2:1 reduction gearing.

118. Hawton, British Gypsum Ltd., Hawton Gypsum Works (SK802506)

Photo taken/Reference: 9.4.1973/4599
Type: Horizontal cross compound
Maker and Date: Marshall Sons & Co. Ltd., Gainsborough. No. 29050. 1898
Cylinder dimensions: 13ins/25ins x 32ins. Drop inlet valves; Corliss exhaust valves
Hp: 158 ihp (1965)* **Rpm:** 44 (1965) **Psi:** 150
Service: Driving grinding and crushing machinery (by belt from twin flywheels)
*The 1965 test reveals that one of the LP inlet valves was suffering from a delay in opening, and that nearer 200 ihp would be obtainable if this was corrected

The engine was delivered new to Cafferata & Co. (predecessors of British Gypsum) in March 1898. The company had been founded by William Cafferata in 1858 to manufacture plaster from gypsum, and a later expansion resulted in the setting-up of the mill at Hawton in the 1890s. This was Marshall's earlier drop valve type – the G Class, with "Proell's two-valve releasing gear" and Corliss exhausts, in this example the GHC variant, cross compound with horizontal air pump and jet condenser behind the LP cylinder. With the valves driven by eccentric rods from the crankshaft, it may be contrasted with the L Class engine, in which the valves are driven from a layshaft (see Wilkin's Tiptree engine – No. 61). The illustration is of the HP side. As will be seen, the engine is extremely dusty. It had stopped work in 1970 and, sadly, was scrapped in 1974.

117

118

119. Kirkby-in-Ashfield, Langton Colliery – No.8 Shaft (SK474551)

Photo taken/Reference: 9.2.1975/25-9
Type: Horizontal twin cylinder winder
Maker and Date: Robey & Co. Ltd., Lincoln. No. 33645-6. 1914
Cylinder dimensions: 21.9ins x 40ins. Drop valves
Rpm: 90.5 (max – 1975) **Psi:** 100
Service: Winding men & materials, depth 630ft. Parallel drum, 10ft diameter

A standard Robey winder, this had first been sold to Cribbwr Fawr Collieries Ltd., Port Talbot, and was moved to Langton to take the place of an earlier engine in the same engine house – possibly in 1935. The cylinders may have been replacements (the NCB figure for their diameter was 21.9 ins compared with the original 24ins shown in Robey's records). Presumably a Richardson governor had been fitted but this and the trip gear mechanism had been removed. When seen, just before its impending replacement by an electric winder, the steam pressure was down to c.50 psi (only one Lancashire boiler out of seven was actually in use), and by early in 1976 both this and the Daglish engine (see *Stationary Steam Engines in Great Britain*, Volume 5, No. 113) had been scrapped.

120. Mansfield Woodhouse, Sherwood Colliery – No.2 Shaft (SK537625)

Photo taken/Reference: 20.3.1977/55-18
Type: Horizontal twin cylinder winder
Maker and Date: Fraser & Chalmers Ltd., Erith. O/No. 6905. 1902
Cylinder dimensions: 26ins x 4ft 6ins. Corliss valves; Allan link reversing
Rpm: 87 (des. max) **Psi:** 150
Service: Winding men and materials from depth 440 yards. Parallel drum, 11ft diameter

This colliery dates from 1902–3 when the shafts were sunk by the Sherwood Colliery Company and two Fraser & Chalmers winding engines were installed, the other being a (larger) cross compound. According to a contemporary *Colliery Guardian* report, the twin cylinder engine was in use in mid 1902 for sinking the shaft, and (improbably perhaps?) would eventually be used with a condensing plant. It was fitted with Seymour patent auxiliary cut-off gear, driven by a third eccentric rod on each side. This was controlled by a Porter governor and was stated to give a cut-off varying from zero to 95 per cent of stroke. By 1977 this had become ineffective because of the speed limits imposed by the overwind settings. The normal rope speed was down to 30ft/sec compared with the 50ft/sec maximum. The original Whitmore controller had as usual by then been replaced by a Black's. Steam winding at the colliery came to an end in 1983 and the engine was immediately scrapped. The No.1 Shaft engine had gone in 1982.

121. Nottingham, Gamble's Factory, Newdigate Street (SK562404)

Photo taken/Reference: 24.11.1980/240-12
Type: Horizontal single cylinder
Maker and Date: Hick, Hargreaves & Co., Bolton. No. 303. 1873
Cylinder dimensions: 13ins x 3ft. Corliss valves
Hp: 50 **Rpm:** 66 1/2 (des.)
Flywheel diameter: 10ft
Service: Driving factory machinery, by belt from 40 inch diameter pulley to lineshaft

This engine was built for James Gamble, a manufacturer of the inside parts of lace machinery – particularly bobbins and carriages for Levers machines, but presumably also supplied power to the lace makers and other occupants of the factory. Hick, Hargreaves engines with Corliss inlet valves operated by Inglis & Spencer's double clip detaching gear were first being built around 1866. This early example, in which the trip gear was controlled by a Porter governor, had a single wrist plate for operating all the valve rods. Later Hick engines of this type used two eccentrics and separate wrist plates for the inlet and exhaust valves, so that they could work independently, making possible greater variation of the cut-off in response to changes in the load. (The engine at Chauntry Mills, Haverhill was of this design – see No.133.) The (8 inch wide) belt drive to the lineshaft remained, but there was no information on working conditions or machinery driven. The engine had been disused since possibly c.1950, when apparently the works had been in the hands of B. Toone (Nottingham) Ltd., hosiery machine manufacturers, for many years. It was removed to the Forncett Industrial Steam Museum, Norfolk in 1981.

119

120

121

122

122. Ollerton, Ollerton Colliery – No. 2 Shaft (SK662674)

Photo taken/Reference: 31.7.1977/78-25
Type: Horizontal twin cylinder winder
Maker and Date: Markham & Co. Ltd., Chesterfield. 1924
Cylinder dimensions: 36¾ins x 7ft. Drop valves
Psi: 155–160
Service: Winding men and materials, from depth 520 yards. Parallel drum, 20ft diameter

Two similar winding engines were ordered by the Butterley Company for its new colliery at Ollerton in 1919. Both had drop inlet and exhaust valves operated via a wrist plate as shown. With Hartnell governor controlling the trip gear and tail rods with tubular supports, No.2 seen here had been little altered over its 50-year life, apart from a possible cylinder rebore and the substitution of a Black's controller for the original Whitmore. However, the coal winder on No.1 Shaft had had replacement cylinders with drop inlets and piston exhaust valves, and a new drum. No.2 Shaft was electrified, and the steam winder scrapped, in 1979 (No. 1 had gone in the preceding year), and in 1994 the colliery was closed.

OXFORDSHIRE

123. Bloxham, Bloxham Grove Farm Mill (SP458367)

Photo taken/Reference: 19.11.1977/94-25
Type: Horizontal single cylinder
Maker and Date: John Lampitt & Co., Banbury. 1852
Cylinder dimensions: c.6ins x 16ins. Slide valve
Flywheel diameter: 6ft
Service: Drove pair of millstones and hoist

This incomplete engine is a rare survival, assumed to date from 1852, when the mill was built. It was typical of the time, a simple slide valve engine with flat bed and 4-bar crosshead guide. However, the top guide bars are missing, as are the connecting rod, two eccentric rods (including one for the boiler feed pump) and presumably a governor. As will be seen, a pair of bevel gears carried the drive to a vertical shaft directly connected to the upper of the single pair of millstones, the weight of which was supported by two iron columns. Behind the bevel gearing is a 14 inch diameter pulley aligned with a belt slot in the wall, presumably allowing for the possibility of connection to an alternative power source. The boiler had been scrapped many years earlier, but the square stack remained.

124. Henley-on-Thames, W.H. Brakspear & Sons Ltd., Henley Brewery (SU763827)

Photo taken/Reference: 15.4.1985/478-7
Type: Inverted vertical single cylinder
Maker and Date: Probably Marshall Sons & Co. Ltd., Gainsborough. No. 28733. 1897/8
Cylinder dimensions: Possibly 5½ins x 8ins. Slide valve
Flywheel diameter: 41¼ins
Service: Drove barrel hoist

This engine has a plate: THE RILEY MANUFACTURING Co., SODA-WATER & BREWERS' ENGINEERS, 256 SOUTH LAMBETH RD, LONDON S.W. However, its appearance is identical to that of Marshall's bottle frame type, and if the number 28733 shown on a separate plate is a Marshall number, the date would be c.1897/8. Presumably it was installed new to drive the hoist, supplied by the Riley company. The engine had two 18 inch diameter pulleys from which open and crossed belts drove to the pulleys on the hoist shaft above – two loose pulleys and a fast pulley between. Either belt could be moved across from its loose pulley to the fast pulley, enabling the hoist to be driven in either direction. In the 1970s, the engine was out of use, the hoist being driven via the engine crankshaft from an electric motor connected by belt to the flywheel. By 1984, this arrangement had been discontinued, the hoist pulleys removed and the hoist shaft direct-coupled to an electric motor. Soon after this, the engine was moved, and its present location is unknown. (The brewery closed in 2002.)

123

SHROPSHIRE

125. Neen Savage, Arthur Butterworth, Goesland Steam Sawmill (SO668791)

Photo taken/Reference: 22.9.1982/357-33
Type: Horizontal single cylinder
Maker and Date: G.Waller & Son Ltd., Stroud. 1927? Installed (second-hand) c.1970?
Cylinder dimensions: c.7ins x 8ins. Slide valve
Psi: 50
Flywheel diameter: 3ft 9ins
Service: Driving sawbench

Arthur Butterworth was a craftsman and steam enthusiast, making high quality furniture. Additional to a number of engines, preserved but not in use, was this small engine, used intermittently for driving a sawbench in an open-sided shed. Reputedly of 1927 and said to have come from a scrap dealer in Ruyton-XI-Towns, it was a very compact engine, with double web crank, single slipper guide and Pickering governor. Drive was by belt from a pulley on the engine to a second motion shaft direct-coupled to a circular saw; for the feed, there were further belts to a reduction gear drive to a pinion engaging a rack underneath. Steam was provided by a mainly wood-fired vertical boiler of 1946 by The Farrar Boiler Works Ltd., Newark.

SOMERSET

126. Street, C. & J. Clark Ltd., shoe manufacturers (ST484368)

Photo taken/Reference: 12.7.1982/345-21
Type: Horizontal tandem compound
Maker and Date: Cole, Marchent & Morley Ltd., Bradford. No. 8697. 1909
Cylinder dimensions: 12ins/24ins x 3ft. Piston drop valves (Morley's Patent)
Hp: 250 **Rpm:** 100 **Psi:** 120
Flywheel diameter: 12ft
Service: Drove machinery and DC generator by 9 ropes 1½ins diameter from grooved flywheel

This was one of Cole, Marchent & Morley's piston drop valve engines, the first of which had been built in 1904 (an inverted vertical compound for the Durham Street Weaving Company, Belfast). Clark's engine, officially started on 1st January 1910, was a replacement for a much overloaded Petrie. The valves were driven by eccentric rods from a side shaft, and had a Hartnell governor for controlling the HP trip gear. It could be worked with steam extraction between the two cylinders, the LP trip gear being controlled by a pressure governor. There was a horizontal air pump and jet condenser in tandem behind the LP cylinder. Towards the end of its life, it drove only a DC generator and workshop shafting – previously it also provided power for other machines (cutting presses etc). Its use was discontinued c.1950-1, when at the same time a Belliss & Morcom driven DC generator was superseded by a new more powerful Belliss AC set. The tandem compound still remains, preserved, although shoe manufacture ceased in 1992.

STAFFORDSHIRE

127. Burntwood, South Staffordshire Waterworks Company, Maple Brook Pumping Station (SK061108)

Photo taken/Reference: 8.4.1995/768-15
Type: Inverted vertical single cylinder rotative air & water pump
Maker and Date: Maker unknown? 1913
Cylinder dimensions: c.8ins x 8ins. Piston valve
Flywheel diameter: 30ins (two flywheels)
Service: Air pump for No.1 Engine

Two very similar inverted vertical triple expansion engines were installed here – one Galloways Ltd., 1913, the second by Glenfield & Kennedy Ltd., 1920. However, a significant difference was that for their surface condensers, the Glenfield & Kennedy had its own air pump, driven from the IP crosshead, while the Galloways made use of an independent air pump. This took the form of a steam engine driven Edwards air pump, located in the basement below. The pump itself, apparently 15 inches in diameter, was direct driven by a pair of pump rods from the crosshead, passing one each side of the double web crankshaft. The crosshead guide was a single slipper on the main back support. The small water (feed?) pump (not visible in the photograph) was driven by a disc crank on the end of the crankshaft; it carries a brass plate: BAKERS (LEEDS) LTD. ENGINEERS LEEDS, but the significance of this is not clear. The pumps would have stopped working with the Galloways engine in 1972, and are presumably still preserved with it.

124

125

126

127

128. Milford, Stafford Corporation Water Department, Milford Pumping Station (SJ975214)

Photo taken/Reference: 12.9.1973/5-5
Type: Inverted vertical triple expansion
Maker and Date: Combe-Barbour Ltd., Belfast. No. E295. 1911
Cylinder dimensions: 14ins/23ins/36ins x 30ins. Corliss valves
Rpm: 26 **Psi:** 150
Flywheel diameter: 11ft 6ins
Service: Water supply – pumping from well 40ft deep to reservoir on Satnall Hills to the south. Normal capacity 50,000 gallons/hour

Milford Pumping Station was officially opened in 1890, the first engines being two horizontals by J. Warner & Sons, London. The Combe-Barbour was housed in an extension of 1912. The Corliss valves were operated by eccentrics on the main crankshaft via intermediate rocking levers, and the HP trip gear was controlled by a chain-driven governor. Three plunger pumps were driven from the crossheads, each with a rod passing either side of the crankshaft. When seen, the engine had been out of use probably for nearly 20 years, and it was scrapped shortly afterwards.

129. Tunstall, Chatterley Whitfield Colliery (SJ884533)

Photo taken/Reference: 20.7.1988/588-3
Type: Horizontal twin cylinder geared winder
Maker and Date: John Warner, Hanley
Cylinder dimensions: 14ins x 22ins. Slide valves; Stephenson link reversing
Psi: 50?
Flywheel diameter: 5ft
Service: At Silverdale Colliery, latterly, until c.1975:
(1) Larger drum used occasionally to pull chains through the rising main in the No.16 upcast shaft, acting as scrapers to remove ochre and corrosion.
(2) Smaller drum used to haul ash cans on to a tip

The engine may date from the 1870s, when the Colliery was undergoing development, and was presumably that recorded in the Silverdale Company's inventory of c.1903 as a pair of hauling engines "in pit No. 15 ... with two round rope drums 9ft and 4ft 6ins diameter and two 6ft twin pulleys". The cylinders were side-by-side with outside slide valves, and the connecting rods forked onto crossheads with two-bar guides. Gearing to the larger drum was by single reduction and to the smaller by double reduction. The photograph is of the engine after its removal for preservation in 1985/6 and re-erection at Chatterley Whitfield.

130. Walsall, John Shannon & Sons Ltd., clothing manufacturers, George Street (SP013983)

Photo taken/Reference: 7.4.1995/766-10
Type: Inverted vertical single cylinder (enclosed)
Maker and Date: Bumsted & Chandler Ltd., Hednesford. No. 1687
Cylinder dimensions: c.6ins x 7ins? Piston valve
Flywheel diameter: 22¾ins
Service: Belt drive to fan for air conditioning plant

The engine and fan, located in the basement of the factory, appeared to have been out of use for many years - the 8-bladed propeller type fan had been removed at some time in the period 1982–91, but the circular recess in the brickwork remained. John Shannon & Son's clothing factory was erected in the period 1894–7, and this was presumably what was left of the air conditioning plant fitted at the time. It was, according to the account in the *Proceedings of the Institution of Mechanical Engineers*, July 1897, a warming and ventilation system "on the plenum plan ... carried out by Mr W. Key, of Glasgow. The external air is drawn through a series of moistened screens, by which it is filtered. It then passes, when warmth is required, through a series of pipes filled with exhaust steam; and is forced through the building by a fan 10 feet diameter, driven by one of Messrs G.E. Belliss and Co.'s compound inverted engines of 12 horse power". The volume of air dealt with in summer was 4½ million cubic feet per hour. This remaining engine, presumably a replacement, was, according to the wording cast on the crank case, Bumsted & Chandler's 7DA model, possibly relating to the cylinder size, and signifying that it was double-acting rather than the company's single-acting type. A shaft governor was linked to a throttle in the steam pipe and the flywheel incorporated a 16 inch diameter belt pulley for driving the fan.

128

130

129

SUFFOLK

131. Brandon, Calders Ltd., timber merchants (TL781873)

Photo taken/Reference: 29.6.1980/222-7
Type: Horizontal single cylinder rotative vacuum pump
Maker and Date: Frank Pearn & Co. Ltd., West Gorton, Manchester. Installed (second-hand) 1948
Cylinder dimensions: c.7ins x probably 8ins. Slide valve
Psi: 90
Flywheel diameter: 3ft
Service: Providing vacuum for creosoting tank

Calders' timber creosoting plant was possibly the last to rely solely on reciprocating steam pumps, which continued in use until the yard was closed and demolished around 1988–9. The timber impregnation was carried out in one of two cylindrical vessels (or autoclaves) in which the creosote is forced into the timber by means of a pressure pump, after which unused creosote is drained into a sump, removal of the excess being completed by a vacuum pump. After this it was pumped back into a storage tank above. (For an account of the site and the process see Chris Allen, "Steam Preservation", *SERG Bulletin*, Vol. 6 No. 1, 1984.) In all there were five horizontal pumps – non-rotative except for the Pearn illustrated: (1) Two Worthington Simpson duplex (one out of use) used as pressure pumps; (2) Two vacuum pumps – the Pearn and another by Joseph Evans, single cylinder; (3) A Lee Howl single cylinder pump used to lift creosote from the sump. Three of the pumps (including the Pearn) had been moved from Calders' premises in Middlesbrough, which had been bombed out in 1942, but they were not brought into use until around 1946–8. The two others were installed later. The Pearn was in a separate corrugated iron shed across the end of the autoclave to which it was connected, but which was little used in the 1980s (most of the work being done in the other). Its pump cylinder, c.8ins diameter, in tandem behind the steam cylinder, had a slide valve with its own eccentric. It is now preserved, not far away from Brandon, at Fengate Farm, Weeting.

132. Brantham, Storey Brothers & Co. Ltd., industrial plastics manufacturer, Cattawade (TM107327)

Photo taken/Reference: 30.11.1984/475-8
Type: Horizontal double tandem compound rotative pumping engine
Maker and Date: Henry Berry & Co. Ltd., Leeds. No. 6528. 1908
Cylinder dimensions: 11ins/17ins x 20ins. Slide valves
Rpm: Up to c.35 **Psi:** 120/170
Flywheel diameter: 7ft
Service: Pumping to accumulator to provide hydraulic power for the factory's presses

This engine was installed here new by The British Xylonite Company, which was formed in 1877 to take over an existing works manufacturing celluloid (xylonite) at Homerton, London. (Xylonite manufacture had been moved to this new factory at Cattawade in 1887.) Providing power for the company's hydraulic presses (by Henry Berry and others) for pressing xylonite, it pumped to a 100 ton accumulator at a pressure of 2,240 psi. After 40 years, it was superseded as the main power provider when a Belliss & Morcom driven 3-throw pump was installed, but it remained in use as a standby. With the closure of a large section of the works in the early 1990s, possibly the last example of the use of reciprocating steam for the production of hydraulic power came to an end. This was an interesting design in which the connecting rods on each side forked onto crossheads which had single slide bar guides above and were located between opposed 2 inch diameter ram pumps. Behind these are the LP cylinders with twin piston rods, and HP cylinder at the rear. Speed was regulated by a Pickering governor, with another throttle valve linked to the accumulator. There were five Babcock & Wilcox boilers supplying steam at 250 psi, a pressure which was reduced for the engine.

131

132

133. Haverhill, D. Gurteen & Sons Ltd., clothing manufacturers, Chauntry Mills (TL672454)

Photo taken/Reference: 17.6.1989/623-19
Type: Horizontal single cylinder
Maker and Date: Hick, Hargreaves & Co., Bolton. No. 519. 1879
Cylinder dimensions: 18ins x 4ft. Corliss valves
Hp: 120 **Rpm:** 55 (des.) **Psi:** 80
Flywheel diameter: 16ft
Service: Drove 150 looms, particularly for the weaving of drabbet

Hick, Hargreaves was one of the first builders of Corliss valve engines in the U.K., using Inglis & Spencer's trip gear. A replacement of two 1856 beam engines by Headley & Manning of Cambridge, this early example, here since new, apparently little altered and still on site, is an extremely important survival. The governor controlling the trip gear was Hick's standard type, and there were separate eccentrics and wrist plates for steam and exhaust valves. (It may be compared with the slightly earlier single eccentric engine at Gamble's Factory, Nottingham – see No.121.) The engine was without a condenser, although it seems that provision for one was made in the foundations. The flywheel, in a separate rope alley, was grooved for 4 ropes, with two ropes driving each way, in opposite directions, to the two lineshafts. The engine stopped work in 1961 when two electric motors driving to the lineshaft pulleys were installed. Weaving ended in 1972, and all manufacturing in 1995, but the company continues to trade and to preserve the engine.

134. Ipswich, Tollemache & Cobbold Breweries Ltd., Cliff Brewery (TM171432)

Photo taken/Reference: 3.4.1991/667-17
Type: Horizontal tandem compound
Maker and Date: E.R. & F. Turner Ltd., Ipswich. No.1967. Possibly c.1900
Cylinder dimensions: 9ins/15ins x 16ins. Slide valves; HP cylinder with fixed expansion slide valve
Rpm: c.80?
Flywheel diameter: 7ft
Service: Drove mash tun rakes, mill(s), and other brewery machinery

This was an important survival, possibly the only engine by this maker now remaining in situ. It was probably installed just after Cobbold & Company's complete rebuild of the brewery in 1894-6. (The maker's name is cast on to the engine as E.R. & F. Turner Ltd., and the company was only registered as such in 1897.) With a girder/trunk guide frame, this was a standard E.R. & F. Turner design, with one eccentric rod to operate the main HP and LP slide valves and a second to drive the HP expansion valve. Speed control was by a Pickering governor, and the air pump/jet condenser was in tandem behind the LP cylinder. Drive was from a 3ft 4ins belt pulley to a second motion shaft running across the engine just over the LP cylinder, and from which a further belt drove to the floors above. It had been out of use for many years when seen, having stopped in 1953 after the East Anglian flood contaminated the well. The engine was restored to intermittent working order in 1979, but the present position is uncertain. Brewing operations, having been transferred to a management buyout company and a new plant, ceased in 2002, but the old brewery may remain intact.

135. Newmarket, Ely, Mildenhall & Newmarket Water Board, Southfields Pumping Station (TL620639)

Photo taken/Reference: 13.1.1974/7-15
Type: Inverted vertical triple expansion pumping engine
Maker and Date: Hathorn, Davey & Co. Ltd., Leeds. 1910
Cylinder dimensions: 12ins/20ins/31ins x 30ins. Corliss valves
Hp: 110 **Rpm:** c.30 **Psi:** 160
Flywheel diameter: 7ft 6ins (two)
Service: Pumping water from well to Newmarket town reservoir. Three ram pumps, 10ins dia x 30ins stroke, pumping against 350ft head

A falling water level and increased demand led the Newmarket Waterworks Company to install this engine as (for most purposes) a replacement of the S.S. Stott beam engine described in *Stationary Steam Engines of Great Britain,* Volume 9 – see No. 120. New engine and boiler houses were erected, with two H. & T. Danks (Netherton) Ltd. Cornish boilers and superheaters. The pumping engine was Hathorn, Davey's standard type of the period, with Corliss valves operated by a high-level camshaft and Craig's trip gear, which for the HP cylinder was controlled by a modified Proell-type governor. Each of the three ram pumps was driven from the crossheads by four vertical rods and there were two flywheels separated by the central IP crank. The design capacity was 1,000 gallons/minute (or c.55,000 gallons/hour), which, however, was apparently never achieved. Electrically-driven pumps were installed in the early 1930s, but steam pumping continued until probably 1964. The engine remained in situ until its removal in 1978 for preservation at Kew Bridge Pumping Station.

133

134

135

SURREY

136. Croydon, Croydon Corporation Water Undertaking, Addington Pumping Station (TQ371628)

Photo taken/Reference: 16.11.1974/23-18
Type: Vertical single cylinder
Maker and Date: Easton & Anderson, London & Erith. 1888
Cylinder dimensions: 5½ins x 9ins. Slide valve
Flywheel diameter: 22ins
Service: Driving overhead travelling crane

The two beam engines in the one engine house are illustrated in *Stationary Steam Engines of Great Britain*, Volume 8 (see No. 118). In addition to the first of these, Easton & Anderson supplied three Cornish boilers and this small vertical engine to drive the overhead crane. Fixed to the wall in one corner, this was a simple but attractive engine, with cylinder and valve chest lagged with wood strips and a single slipper guide fixed to the plate against the wall. The crane shaft, 2ins square, was connected by a vertical belt to the engine flywheel/pulley, and ran along the length of the engine house. The crane was operated by a bevel gear sliding along the square shaft, the supports of which pivot out of the way to allow the gear to pass. The engine remained in occasional use until 1979 at least (some years after the beam engines were stopped), but was finally removed for preservation (with the crane) in 1983.

137. Croydon, Croydon Corporation Water Undertaking, Waddon Pumping Station (TQ313639)

Photo taken/Reference: 2.8.1981/293-11
Type: Cornish boilers
Maker and Date: Ruston, Proctor & Co. Ltd., Lincoln. 1909
Psi: 100
Service: Providing steam for two horizontal cross compound pumping engines

By the time pumping finally stopped in 1983, Waddon was the last waterworks in Great Britain using reciprocating steam engines, and at least one of the original three 6ft x 20ft Cornish boilers, still hand-fired, remained in use. They had Thomas Beeley & Son superheaters and were also originally equipped with underfeed mechanical stokers, but the latter were removed many years ago, presumably when the boilers were fitted with Halliwell furnaces. In later years, the working pressure had been reduced from the original 120 psi to the present 100 psi (at 440°F). Around mid 1981, the brickwork of the No. 2 (middle) boiler flue wall was found to be collapsing and considered too difficult to repair. Although a single boiler was sufficient for one engine, this seems to have helped to precipitate the final decision to replace the engines with electric pumps. (The main pumping engines are described in *Stationary Steam Engines of Great Britain*, Volume 8 – see No. 120.)

138. Croydon, Croydon Corporation Water Undertaking, Waddon Pumping Station (TQ313639)

Photo taken/Reference: 23.3.1983/385-9
Type: Horizontal single cylinder
Maker and Date: E. Green & Son Ltd., Wakefield. 1910? (An insurance inspection report gives 1935!)
Cylinder dimensions: 3½ins (prob.) x 6ins. Slide valve
Rpm: c.180
Flywheel diameter: 18ins
Service: Driving economiser scrapers

The boiler plant put in at Waddon in 1910 included a 64 tube Green's fuel economiser. The subject of a patent by Edward Green in 1845, this enabled the boiler feed water, passing through vertical pipes, to be pre-heated by waste flue gases from the boilers. To keep the outsides of the pipes free of soot, slow-moving scrapers are operated by chains (in this case four), driven by worm gearing from a longitudinal shaft, the rotation of which is automatically reversed by driving it alternately by one or other of two facing bevel wheels. The scraper gear could be driven by an electric motor or as here by one of Green's standard engines, designed for the purpose. A slide valve engine, with trunk guides, overhung cylinder and Pickering type governor, this was the smaller of the two sizes made. The drive could be by belt from the flywheel, but in this case was from an 8 inch diameter pulley, with the belt stretching from one end of the economiser to the other. The usual practice here was not to operate the scrapers continuously but once per shift, in contrast with the maker's manual instruction "to keep the Scrapers constantly running when the Economiser is in use"! The engine was removed for preservation soon after steam pumping ceased in 1983, but its present location is unknown.

137

138

SUSSEX

139. Brede, Hastings Waterworks Department, Brede Pumping Station (TQ814178)

Photo taken/Reference: 13.7.1973/3-7
Type: Inverted vertical triple expansion pumping
Maker and Date: Tangyes Ltd., Birmingham. 1904
Cylinder dimensions: 23¼ins/38ins/62ins x 4ft. Slide valves; Meyer expansion valves (HP & IP)
Hp: 410 **Rpm:** 23 (max) **Psi:** 150
Flywheel diameter: 15ft
Service: Pumping to Fairlight Reservoir for the water supply of Hastings. Three high lift ram pumps, 12½ins x 4ft. Lift 495ft, capacity 80,000 gallons/hour

A number of the earlier waterworks engines of this type had slide valves, and it is fortunate that two at least survive (the other is at Hereford – see *Stationary Steam Engines of Great Britain*, Volume 6, No. 95). This was one of a pair installed to pump from two wells sunk to a depth of 275ft in 1903 (the other was scrapped in 1969). However, the well pumps, two bucket & plungers, 18½ins x 3ft, driven from the HP end via opposed bell cranks, were disconnected in 1933, having been superseded by electrically-driven pumps. This left the engines to drive only the high-lift pumps, driven from the crossheads, pumping water raised from the wells and from other sources (well supplies were insufficient). A Tangye governor linked to a throttle valve controlled the engine speed, and exhaust was to a surface condenser. This had a horizontal air pump at floor level driven by a disc crank at the LP end (visible in the illustration). When the new Worthington-Simpson engine (see No. 140) was installed in 1939–41, the original four Babcock & Wilcox water tube boilers were replaced by new ones – also Babcock & Wilcox, but higher pressure and with superheat. For the Tangyes, which had used saturated steam and continued in use until 1964, and as standby until the 1970s, the steam had to be "de-superheated".

140. Brede, Hastings Waterworks Department, Brede Pumping Station (TQ814178)

Photo taken/Reference: 22.5.1977/72-11
Type: Inverted vertical triple expansion pumping
Maker and Date: Worthington-Simpson Ltd., Newark. No. 5058. 1939
Cylinder dimensions: 17ins/31ins/56ins x 4ft. Piston drop valves
Hp: 420 **Rpm:** 27 **Psi:** 250
Flywheel diameter: 13ft (two)
Service: Pumping to Fairlight Reservoir (lift 515ft) by 16¾ins ram pump; and to Baldslow Reservoir (lift 400ft) by two 13ins ram pumps. Capacity c.3½ million gallons/day at 30 rpm

This was the last engine of this type ordered from Worthington-Simpson, although one of the two slightly smaller Dover Waterworks engines (No. 5056 of 1937) was not actually commissioned until 1954 (see *Stationary Steam Engines of Great Britain*, Volume 8, No. 13). It was installed to increase the high lift capacity in a new engine house, the pumps being driven from the crossheads, the 16¾ins ram from the HP, the two 13ins rams from the IP and LP. A high level eccentric shaft operated the valves, with the HP trip gear governor-controlled. After electric pumps took over in 1964, the steam engines remained on standby into the 1970s – the No.2 Tangyes was steamed for the last time in November 1975. Soon after this, the Babcock & Wilcox boilers of 1939 (two) and 1948 were scrapped, but both remaining engines are being preserved.

141. Horsham, Christ's Hospital – laundry (TQ150290)

Photo taken/Reference: 3.8.1981/295-22
Type: Horizontal single cylinder
Maker and Date: Robey & Co. Ltd., Lincoln. No. 21632. 1901
Cylinder dimensions: 9ins? x 18ins. Slide valve
Rpm: c.115? **Psi:** 100?
Flywheel diameter: 5ft 6ins
Service: Drove laundry machinery

This was a Robey "A" class engine, delivered from the works to W. Summerscales & Sons Ltd., Phoenix Foundry, Keighley on 27 January 1902. This date corresponds with that of the establishment of the school at Horsham and therefore presumably that of the laundry, which suggests that the engine may have been installed here new. It was a simple slide valve engine with bayonet/trunk guide frame, supported on separate beds for the main bearing and cylinder. Steam was admitted via a Pickering governor-controlled throttle valve and almost certainly exhausted to the laundry. Driven via a lineshaft from a 3ft diameter belt pulley was the usual laundry machinery – washing and ironing machines, and a compressor. Now, following a modernisation around 1969–70, when presumably the engine was stopped, all that was left of this was a calender. Electrified and in a different position, this was a rotary steam heated iron for sheets etc., supplied in 1947 by D. & J. Tullis Ltd. of Clydebank. The engine was removed for preservation soon afterwards and is now at Chatham Dockyard.

139

140

141

142. Hove, Goldstone Pumping Station (now Brighton & Hove Engineerium) (TQ286066)

Photo taken/Reference: 15.4.1992/683-33
Type: Horizontal single cylinder
Maker and Date: Reputedly Easton, Amos & Sons, London. 1862
Cylinder dimensions: c.10ins x 18ins. Slide valve
Flywheel diameter: 7ft 2ins
Service: Drove waterworks workshop machinery

The workshop engine, apparently installed in 1875, was reputedly built by Easton, Amos & Sons for the International Exhibition of 1862. There may be no evidence for this, but its design, with flat bed, 4-bar guide and early Watt-type governor, seems to fit the date. It presumably continued in use, driving the workshop lineshaft from a 4ft 4ins diameter belt pulley, until steam pumping ceased in the 1950s (the earlier of the two main engines is dealt with in *Stationary Steam Engines of Great Britain*, Volume 8 – see No. 137). When first seen in 1972, surrounded by mesh fencing, it looked very much a working engine, but now, preserved as one of the Engineerium's exhibits, it is shown here much restored and with its cylinder lagging removed.

143. Hove, Brighton Corporation Water Department, Hove "Sea-Water Pump", King's Esplanade (TQ290043)

Photo taken/Reference: 15.4.1992/683-12
Type: Inverted vertical single cylinder rotative pump
Maker and Date: Eastons & Anderson, London. 1872
Cylinder dimensions: 9ins x 9ins. Slide valve
Flywheel diameter: 4ft
Service: Supplied seawater for bathing to houses and hotels

This was a "Scotch crank" pump, initially operated by compressed air, installed by The West Brighton Waterworks Company Ltd. to supply seawater to local houses and hotels. Located in an underground chamber about 20ft below Hove seafront, it is accessed through a manhole and by descending a vertical ladder. The slide valve cylinder is supported on a double A-frame which also incorporates the crosshead guides. Below, direct-driven, is the $4^{1}/_{2}$ inch diameter bucket and plunger pump, delivering water initially to the air vessel which can be seen in front of the A-frame. The sea-water could be pumped into or from underground storage tanks or direct to houses from the sea. It could also flow from the sea into the tanks at times of high tide. In 1876, the West Brighton company was purchased by Brighton Corporation, and the pump was converted to hydraulic operation using mains water. To prevent hydraulic locking at the end of each stroke, resulting from the lap of the valve (water being non-compressible), the ends of the cylinder were connected to the supply main via small air vessels with non-return valves. The pump apparently continued to operate until 1940, when it became inaccessible because of wartime beach defence installations.

142

143

144. Lewes, Beard & Co.(Lewes) Ltd., Star Lane Brewery, 12 Fisher Street (TQ414102)

Photo taken/Reference: 8.8.1973/4-3
Type: Inverted vertical single cylinder
Maker and Date: A. Shaw, Lewes. c.1880–1900
Cylinder dimensions: c.7ins x 1ft. Slide valve
Psi: 60
Flywheel diameter: 3ft 11ins
Service: Drove brewery machinery

Presumably installed new in a brewery which had been founded in 1741 and acquired by Thomas Beard in 1811, this was a basic engine of almost the simplest design possible, built by the local millwright Alfred Shaw, who is known to have flourished around 1878–1903. Nevertheless, it was of a slightly unusual design, with the cylinder supported on a hollow column containing the semi-circular slippers of the trunk crosshead guide and slotted to accommodate the connection with the external connecting rod/piston rod linkage. The two main bearings supporting a bent crankshaft were mounted on a cast iron box, with the overhung flywheel at one end. At the other end was a $19^{5}/_{8}$ins belt pulley from which the drive was taken vertically to the second motion shaft above. There seemed to be no provision for a governor. Brewing and the use of the engine ceased in December 1959, when production was transferred to Harveys' Bridge Wharf Brewery, also in Lewes, but the engine remained here until its removal to Harveys in 1985.

WARWICKSHIRE

145. Long Compton, Weston Park Estate sawmill (SP287345)

Photo taken/Reference: 29.7.1978/120-18
Type: Vertical single cylinder
Maker and Date: Unknown. Installed c.1854
Cylinder dimensions: c.12ins x 2ft. Slide valve
Psi: c.60
Flywheel diameter: 9ft
Service: Drove sawmill machinery, corn mill and chaffcutter

The sawmill was started c.1854, at which time the estate was in the hands of the 2nd Earl of Camperdown. The engine was said to have been second-hand, so may have been built as early as the 1840s. An interesting design, maker unknown, it was presumably the product of a local country foundry. An A-frame supported the main bearing and single web overhung crank. Between this and the flywheel against the wall was an eccentric rod operating the slide valve via an intervening lever and, mounted on an iron column, the Watt governor, overhead driven by bevel gearing. Driven by round side rods from the crosshead is a boiler feed pump, the barrel of which is in a casting which incorporates the crosshead guides on each side. The sawmill was driven by leather belt from the flywheel, and two small pulleys beyond the wall (one 29ins diameter remaining) drove the cornmill, but apart from an iron sawtable loose in the yard, the machinery had disappeared. There was a replacement Cornish boiler: Edwin Danks & Company (Oldbury) Ltd. No. 5138 of 1907, maximum working pressure 80 psi. Milling came to an end at about the time Weston Park was demolished in 1933, but the estate and mill remains in the hands of the Warriner family, which had taken over in 1918.

146. Rugby, GEC Machines Ltd., Mill Road (SP510764)

Photo taken/Reference: 18.2.1989/607-19
Type: Inverted vertical compound compressor (enclosed)
Maker and Date: Belliss & Morcom Ltd., Birmingham. No. 1529. 1924
Cylinder dimensions: 20in/29⁷/8ins x 14ins. Piston valves
Rpm: c.85–120 (variable) **Psi:** 110
Flywheel diameter: c.6ft
Service: Supplying works compressed air at 80 psi

This engine was installed new at Blidworth Colliery in Nottinghamshire, where it would have worked at nearer to its design conditions: steam pressure 140 psi; 175 rpm; 538 hp; capacity 3,000 cubic ft/minute. Becoming redundant there, it was purchased by GEC in 1973, and by June 1974 had been installed in their Power House at Rugby. Here, it continued to work during winter months until 1989. After possibly a year or so on standby, it was scrapped with much other machinery, when in 1991 the Power House dating from 1898 was demolished. As will be seen the air cylinders are mounted in tandem above the steam cylinders: 1st stage (29⁷/8ins dia) above the LP; 2nd stage (18⁷/8ins dia) above the HP. The HP piston valve was fitted with hand-adjustable variable expansion gear, and the exhaust steam was passed to the works process main.

WORCESTERSHIRE

147. Beckford, A.J. Woodward & Sons Ltd., market gardeners & poultry farmers (SO981359)

Photo taken/Reference: 8.1.1992/678-4
Type: Inverted vertical single cylinder (enclosed)
Maker and Date: Spillingwerk GmbH, Hamburg. No. 1982. Installed 1990–1(second-hand?)
Cylinder dimensions: 410mm x 126mm. Piston valve
Hp: 200 **Rpm:** 1,000 **Kg/cm²:** 6.0
Flywheel diameter: 750mm
Service: Electricity generation by 160kW alternator

A type of reciprocating steam engine built in quite large numbers and still being installed in recent years in various parts of the world is the Spillingwerk company's high-speed "steam motor" – perhaps mostly in Germany, but a few also in Great Britain and Northern Ireland. First introduced in the 1950s, these are of modular design – self-contained units, which could be coupled (up to five or six) together as required. This example of the Spilling motor was a single unit of the Type 12 size, erected in part as an experimental installation monitored by the Department of Energy to determine the technical and economic viability of a small-scale heat and power plant using poultry litter as the boiler fuel. The engine exhausted at 0.7 kg/cm² to heating coils for heating of the chicken sheds (or to atmosphere), back pressure, engine speed etc. being controllable by a hydraulic governor. The alternator and exciter, direct-coupled, were also by Spillingwerk. (In practice, the experimental fuel was found unsatisfactory and coal was substituted!)

146

147

148. Tenbury Wells, Tenbury Wells Pump Room & Brine Baths (SO596683)

Photo taken/Reference: 17.9.1984/473-17
Type: Inverted vertical single cylinder
Maker and Date: E.S. Hindley & Sons, Bourton. Installed 1911
Cylinder dimensions: 4ins x 6ins. Slide valve
Flywheel diameter: 21ins
Service: Drove pump(s) in well

The saline spring at Tenbury Wells was discovered in 1839, and the strange buildings, roofed with iron sheets, completely derelict by 1984, were erected in 1862 for The Tenbury Wells Improvement Company. They included an octagonal water tower over the well, the lower part of brick, surmounted by a pagoda type roof of which all that was left was the wooden framework containing the two water tanks. The engine was installed at the bottom of the tower in 1911, and although nothing to identify the maker could be seen, it appeared identical to E.S. Hindley's "Alcazar" type. The pumps, at a depth of 58ft, were driven from one end of the crankshaft, possibly via gearing and a 2-throw crank – all missing by 1984. The presence of switches and electric cables suggested that the engine had been superseded by an electric pump. The buildings have now been refurbished, but unfortunately are without the engine, which was removed, probably in 1994, and is now being preserved at Broomy Hill Pumping Station, Hereford.

YORKSHIRE: EAST RIDING

149. Cottingham, Yorkshire Water Authority, Cottingham Pumping Station (TA047342)

Photo taken/Reference: 23.10.1982/381-9
Type: Inverted vertical triple expansion pumping
Maker and Date: Worthington-Simpson Ltd., Newark. No.5048. 1930
Cylinder dimensions: 18ins/31ins/50ins x 4ft. Piston drop valves
Hp: 350 php **Rpm:** 24 **Psi:** 200
Flywheel diameter: 14ft 3ins
Service: Pumping to Keldgate No.1 Reservoir for water supply of Kingston upon Hull. Two well pumps (19ins x 4ft stroke); three ram pumps (13½ins x 4ft). Capacity 5 m.g.d.

Mill Dam Pumping Station, as it was then called, was opened by Kingston upon Hull Corporation Water Department in 1891, with three inverted vertical compound engines by Amos & Smith, of Hull. The engine illustrated here is the last surviving of three Worthington-Simpson engines (Nos. 5046–5048) dating from an extension of 1928–30. (Only five further such engines were built by the Company, the last being the very similar No.5058 of 1939 – see No.140.) Three ram pumps were driven by side rods from the engine crossheads; and the two bucket well pumps were driven off the HP and LP ram pumps. Steam was provided by two Lancashire boilers. The three engines continued in use until 31st May 1974, after which two were scrapped but the third, seen here, was preserved in situ by the Yorkshire Water Authority, which had taken over two months earlier.

150/151. Kingston upon Hull, British Transport Docks Board, Alexandra Dock (TA128289)

Photo taken/Reference: 23.10.1982/379-28 & 380-19
Type: Stationary wharf crane
Maker and Date: James Taylor & Co., Birkenhead. 1886
Cylinder dimensions: (1) Horizontal twin cylinder 12ins x 16ins (for hoisting); (2) Inverted vertical twin cylinder 7ins x 10ins (for slewing). Both with slide valves and Stephenson links
Rpm: 60 (hoisting engine); 80 (slewing engine) **Psi:** 100
Service: Dockside crane. S.W.L. 100 tons

Alexandra Dock, built for the newly constructed Hull & Barnsley Railway, primarily as an outlet for the South Yorkshire coalfield, was opened in 1885. The crane would have been installed a year or two later. The stencilled "No.69" on the crane was an LNER reference, this company having taken over in 1923. The gearing for slewing involved four stages ending up with worm gears to two separate pinions meshing with a circular rack beneath, the crane revolving on a circle of rollers. Hoisting through three pairs of gears could be carried out via either of two ratios, resulting in lifting speeds of 10ft/minute or (for heavier loads) 5ft/minute. The boiler pressure (160 psi) was reduced after a replacement vertical boiler was fitted in 1976. No longer in use for coaling, in 1980 the crane was reported as still occasionally making commercial lifts and also quite frequently lifting various pilot vessels out of the dock. Two years later the Docks Board left Alexandra Dock and over the next decade the crane became increasingly derelict.

152. Kingston upon Hull, Hull Central Dry Dock, Humber Street (TA101282)

Photo taken/Reference: 23.10.1982/378-36
Type: Inverted vertical single cylinder
Maker and Date: Amos & Smith, Hull. No.15?
Cylinder dimensions: c.14ins x 1ft. Slide valve
Psi: 100
Flywheel diameter: 3ft 6ins
Service: Driving Amos & Smith centrifugal pump. Used to pump out dry dock

There were three steam engine driven centrifugal pumps, possibly installed by the Hull Central Dry Dock & Engineering Company, owners (and builders?) of the dock in the nineteenth century and until around the mid 1930s. Subsequently, the dock was successively in the hands of the engineers, Amos & Smith Ltd. until the 1960s; the Drypool Engineering & Dry Dock Co. Ltd. until 1976; and Ruscador Marine & Electronics Ltd. until 1992. The engines were located in two circular wells beside the dock, connected by a passage, in one of them the engine and pump illustrated, in the other a pump set of the same size plus another smaller. All three, engines and pumps, direct-coupled, are assumed to be by Amos & Smith, the name together with the No. 15 cast onto the pump casing of the engine illustrated. It is of a heavy rather utilitarian construction, with double web counterbalanced crank and small overhung flywheel for barring. Exhaust went to atmosphere. Around 1974/5 the other two pumps were electrified, but the steam pump which remained continued in use until at least 1991. In 1992 the Ruscador company was reported to be in receivership and the dock was abandoned, scheduled for development. But the pumps were possibly still in situ at the time of writing.

148

149

150

152

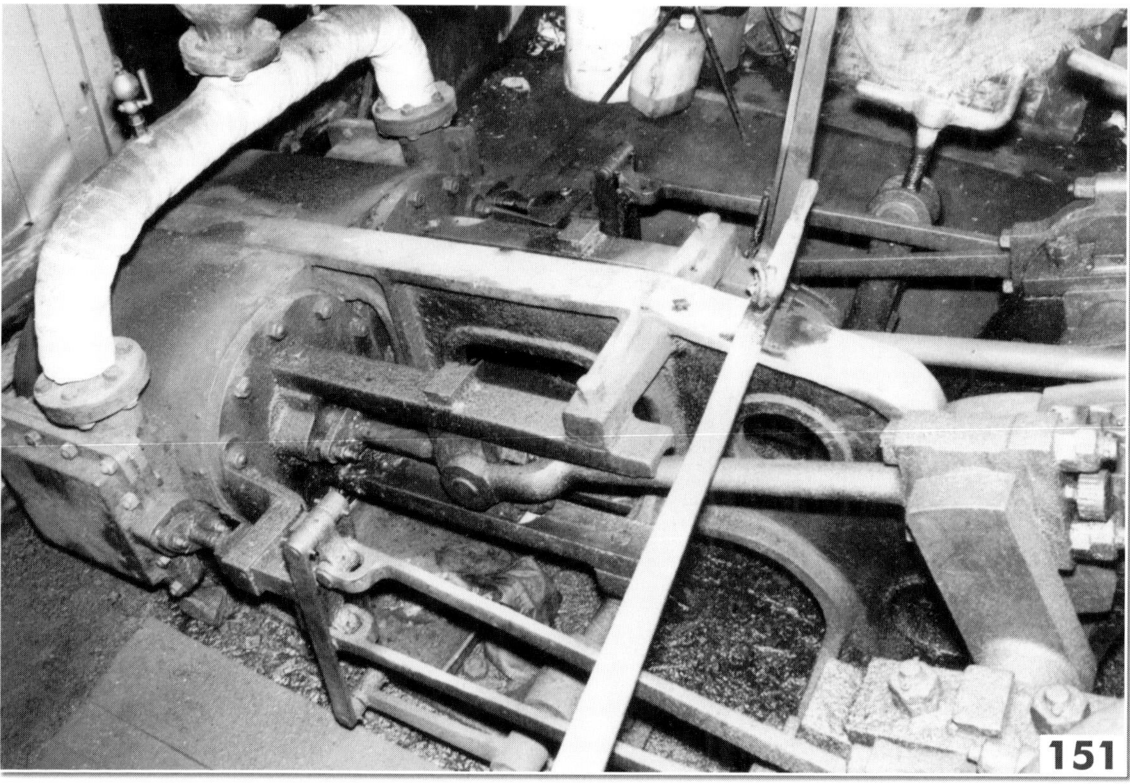

151

153. Newport, Henry Williamson & Co. Ltd., Newport Brick Works (SE853308)

Photo taken/Reference: 26.3.1974/8-14
Type: Horizontal single cylinder
Maker and Date: Robey & Co. Ltd., Lincoln. No.20948. 1900
Cylinder dimensions: 15ins x 33ins. Drop valves; slide exhaust valves
Rpm: 86 **Psi:** 82
Flywheel diameter: 11ft
Service: Driving pipe machine and clay haulage incline

Henry Williamson & Co. Ltd. were part of the Crossley Building Products Group, and this engine, installed here second-hand around 1925, was practically at the end of its working life, which came later the same year. It was another of Robey's E Class drop valve engines, in which the valves were worked by eccentrics on the skew gear-driven side shaft (see also under Fairfield Hospital – No.1; and Sevenoaks Brick Works – No.80). The drop valves were operated by "Richardson & Rowlands Patent Cut Off Gear", but in this example the Richardson governor, rather than being above the cylinder, was set further forward. It was driven from the layshaft via a moveable belt between two cone pulleys, presumably a requirement for its initial use. Robey's records show that it first went to Bentinck West Hartley Colliery Ltd., who were the owners of Pegswood Colliery, near Morpeth. When seen the engine speed was slightly below the (maximum?) design speed of 90 rpm. The drives, initially by belts from two pulleys, were to a pipe extruding machine (the pipes being cut by hand) and to a haulage incline, drawing small trucks of clay from the pit. There was a single coal-fired Lancashire boiler by R. & J. Dempster Ltd., Manchester. Soon after it stopped work, the engine was moved to the Tyne & Wear Museums, Newcastle.

YORKSHIRE: NORTH RIDING

154. Guisborough, The Carrs Tilery (Skelton Estate brickworks), Margrove Park (NZ651160)

Photo taken/Reference: 29.3.1974/9-8
Type: Vertical single cylinder
Maker and Date: Samuel Riddle & Sons, Tweedmouth Forge. 1871
Cylinder dimensions: 15ins* x 2ft. Slide valve
Hp: 20
Flywheel diameter: 11ft
Service: Drove wire-cut brick machine by Henry Clayton Son & Howlett (The diameter is given as 16ins in Riddles' offer to build the engine, but 15ins in the 1894 inventory.)

The development of the Skelton Estate brickworks (owner Mr. J.T. Wharton of Skelton Castle) was particularly associated with the exploitation of the Estate's ironstone deposits from 1863. The Tilery was active in the period 1867–76, producing tiles, pipes and bricks. The installation of the engine was linked to an expansion in the period 1871–2, but a depression in the iron industry resulted in an almost immediate collapse in demand. Certainly, all was stopped by 1894, when an inventory was made of the machinery, which included the gear-driven brick machine and a Lancashire boiler 15ft 6ins x 6ft 6ins. The engine, although not the boiler and brick machine, was largely still intact in the 1960s, although without connecting and eccentric rods and governor. (For a record of the site in 1967 and 1969, with drawings and photograph of the engine, see J.K. Harrison, "The Carrs Tilery, Margrove Park", *The Cleveland Industrial Archaeologist* No. 16, 1984.) Unfortunately, around 1970, the substantial two-columned iron framework supporting the main bearing (the crankshaft was held between this and a bearing on the now much reduced wall to the left) was removed, leaving the present remains. As will be clear, the connecting rod had forked onto the crosshead, and for the piston rod there was a bush guide held by a bracket off the frame. In recent years the site, which includes an unsafe looking square chimney, has become increasingly overgrown.

153

154

YORKSHIRE: WEST RIDING

155. Baildon, Baildon Combing Co., Lower Holme Mills (SE154382)

Photo taken/Reference: 30.4.1979/155-4
Type: Uniflow
Maker and Date: Sulzer Bros., Winterthur. No.4799. 1922
Cylinder dimensions: 850mm x 900mm. Double beat drop valves
Hp: 1,000 **Rpm:** 133 **Psi:** 135
Flywheel diameter: 14ft
Service: Direct drive into mill shed, rope drive to alternator

In the 1950s, this engine, noted for its practically silent operation, had been regularly working a continuous 125-hour week, but by 1979 it was completely derelict; much of the mill had been demolished, with the engine shortly to follow. Typically for this type, it was heavily built, mostly enclosed, including the double web crank. There were three bearings, water-cooled. Positively operated drop valves were driven by two variable eccentrics on a bevel gear driven layshaft, controlled by the shaft governor. The horizontal air pump in the basement was also enclosed, as was the (chain?) drive from the end of the engine shaft. The flywheel incorporated a smaller (9ft) diameter 16-grooved rope drum, from which 7 ropes drove an alternator. A narrower 8ft 10ins diameter section provided for a flat belt to a pump below. Beyond this was a coupling to the main mill lineshaft.

156. Barkisland, Edmund Sykes & Sons Ltd., woollen manufacturers, Krumlin Mill (SE055183)

Photo taken/Reference: 22.8.1973/4-35
Type: Horizontal tandem compound
Maker and Date: Pollit & Wigzell Ltd., Sowerby Bridge. 1899
Cylinder dimensions: $10^1/_4$ins/23ins x 3ft. Corliss valves (HP); slide valve (LP)
Hp: 180 ihp **Rpm:** 85 **Psi:** 150
Flywheel diameter: 12ft
Service: Wool spinning. Rope drive

Edmund Sykes manufactured fancy woollen cloths, having taken over Krumlin Mill in 1914. The Pollit & Wigzell engine was installed 1915–16, replacing a beam engine, and having come here second-hand from The London General Omnibus Company, Bromley. There, it was one a pair used for driving machinery for the preparation of fodder for horses, which became redundant with the introduction of motor transport. Sold to Edmund Sykes & Sons for £120, it was uprated from its original designed 150 ihp at 80 rpm, and in January 1923 fitted with a replacement (Lumb) governor and regulator for controlling the HP trip gear. It was Pollit's patent three piston rod type with the usual arrangement of jet condenser and air pump in tandem behind the LP cylinder. The drive was by 7 ropes $1^5/_8$ins diameter (only 5 on in 1973) from the grooved flywheel. With activity declining, it finally stopped working on 28th February 1975, probably remained on standby for a few years, but was removed for private preservation in 1980.

157. Bradford, Charles Fox & Co. (Dyers) Ltd., Jesse St. Dyeworks (SE163321)

Photo taken/Reference: 21.7.1990/653-27
Type: Horizontal single cylinder
Maker and Date: Unknown. Installed c.1900
Cylinder dimensions: c.15ins x 30ins. Slide valve
Flywheel diameter: 9ft 9ins
Service: Drove hydro-extractors (etc?) – possibly by belt from flywheel rim

The maker and date of this engine are unknown, although Thomas Metcalfe Ltd. of Bradford and W. Bradley of Brighouse have been suggested as possibilities. The building apparently dated from c.1880 as the premises of a wool and waste dealer, and an engine house was added in 1886. It was in use as a dyeworks in 1895, but the installation of this engine must have dated from c.1900 when the works were re-modelled with a replacement, re-located engine house – perhaps linked with the establishment here of Charles Fox & Co. This was a slide valve non-condensing engine of the most basic type, with single slipper crosshead guide, disc crank and Pickering type governor. A second eccentric on the crankshaft seems not to have had any use, and exhaust presumably went to process. When seen, the works had been closed for some time, and the machinery had been cleared, although there was still a relatively modern boiler. Also remaining were the two halves of a 5ft 3ins diameter belt pulley, which could have come off the engine. Removed in 1991, the engine has now been installed for preservation in Quarry Bank Mill, Styal.

155

156

158. Bradford, Holmes, Mann & Co. Ltd., packing case manufacturers, Harris Street Sawmills (SE172333)

Photo taken/Reference: 2.6.1972/4076
Type: Horizontal single cylinder
Maker and Date: Newton, Bean & Mitchell, Dudley Hill, Bradford. c.1910?
Cylinder dimensions: 17ins x 30ins. Corliss valves
Hp: 100 **Rpm:** 100 **Psi:** 120
Flywheel diameter: 12ft
Service: Electricity generation. Belt drive from flywheel to English Electric Co. Ltd. alternator. 125 kVA

Between 1896 and 1936, Newton, Bean & Mitchell built a number of mainly medium-sized mill engines, two of which were installed here in what was then the works of woolcombers, H.D. Shaw & Co. One was the horizontal cross compound drop valve engine installed here new in 1921 and described by George Watkins (see *The Textile Mill Engine*, 2nd Edition, 1999). The other was this smaller non-condensing Corliss valve engine said to have been put in at the same time but second-hand and dating from possibly c.1910. (No definite information about its earlier history has come to hand.) Neatly built, the squared off shape of the Corliss valve casing seems to be typical of Newton, Bean & Mitchell, as was the use of William Eastwood's isochronous "Bee" governor for controlling the trip gear – also present on the cross compound. It was still giving good service after more than 50 years use, but both engines were scrapped in 1975.

159. Bradford, J.C. Thompson & Co. (Duron) Ltd., textile chemical manufacturers, Duron Works, Drummond Road (SE150342)

Photo taken/Reference: 30.4.1979/153-35
Type: Horizontal single cylinder
Maker and Date: Unknown. Possibly installed c.1915–20
Cylinder dimensions: c.8ins x 1ft. Slide valve
Psi: 75–80 **Flywheel diameter:** 4ft
Service: Driving paddles in two mixing vats for manufacture of textile chemical dyes

The British Duron Company, oil emulsion manufacturers (previously at 3 Vaughan Street as The Duron Co. Ltd.) seem to have established the Duron Works here around 1915–20. The company was taken over by Mr. John Clifford Thompson in 1930. Little seems to be known for certain of the history of this engine. Possibly it was installed when the works was established (i.e. around 1915–20), although a 1905–6 installation date has been given. Thomas Bradford & Co., Manchester has been suggested as a possible maker, based on design similarities. With a hollowed out flat bed, single slipper crosshead guide and Pickering type governor, this was a basic design. The drive was to a lineshaft by belt from a 2ft diameter pulley. The only features of note were a rather unsightly support for an oil cup and a small boiler feed pump driven by a forward extension of the valve rod. The engine ceased working shortly after being seen, possibly later in the same year, but was not removed (to Farling Top Stationary Steam Engine Collection, Cowling, Yorkshire) until 1993.

160. Castleford, Whitwood Sewage Pumping Station, Cinder Lane (SE422259)

Photo taken/Reference: 20.2.1982/314-8
Type: Two horizontal single cylinder pumping engines (workable singly or coupled) (remains)
Maker and Date: Joshua Horne, Providence Ironworks, Castleford. 1882
Cylinder dimensions: Cylinders (12½ins x 2ft. Slide valve; Meyer expansion valve) removed
Rpm: (18) **Psi:** (c.50) **Flywheel diameter:** 6ft each
Service: Pumping sewage through a 2,300 yard main with a rise of 68ft to the sewage farm, from which after purification it passed to the River Calder

The sewerage system for Whitwood UC designed by the consultant, Mr John Richardson, included this steam pumping station located at the lowest level of the Whitwood Mere district. Drawings of the engines in their original form are provided by a contemporary account – see "Whitwood Sewerage Works", *The Engineer*, Vol. LV, 1883. Completely derelict when seen, the plant remained as last used, converted to electric driving. In this form, original engine beds, pumps, crankshaft and flywheels were retained, but much had been removed, including steam cylinders (replaced by the single slipper guides) and, at the end opposite the sewage pumps, the air pump/condensers. Originally, banjo type linkages passed round the overhung cranks on each side, connecting piston and air pump rods. The double-acting pumps (14ins x 2ft stroke) pumped 30,000 gallons/hour, which could, it was, said be doubled if required. Two Cornish boilers, working pressure 50 psi, were installed in 1882, but these were presumably superseded by refuse destructor boilers, not seen, but believed scrapped c.December 1981. After the conversion, the pumps were operated by the original crankshaft, driven by herringbone reduction gearing (c.5:1) from a newly inserted shaft, belt driven by an induction motor (English Electric Co. Ltd., London). Unfortunately, the dates of conversion and final stoppage are not to hand, and all was scrapped probably within a year or so of the visit.

161. Dewsbury, William Greenwood & Son Ltd., woollen manufacturers, Providence Mill, Earlsheaton (SE257213)

Photo taken/Reference: 20.8.1984/458-26
Type: Horizontal tandem compound
Maker and Date: W. & J. Cardwell, Dewsbury. 1883; Woodhouse & Mitchell, Clifton Bridge Iron Works, Brighouse. O/No.491. 1894
Cylinder dimensions: c.15ins/c.20ins x 42ins. Corliss valves (HP); slide valve (LP)
Hp: c.250 **Rpm:** 85 **Psi:** 120? **Flywheel diameter:** 14ft
Service: (Previously) driving woollen spinning machinery etc.
(Originally a single cylinder engine, in 1894 compounded by the addition of the HP cylinder by Woodhouse & Mitchell.)

This was built for William Greenwood & Son in 1883 (although the name of the maker on the slide valve cover, W. & J. Cardwell, was only revealed in 1985 during restoration work). In 1894, the mill was rebuilt and extended after a fire, and the increased power requirement resulted in the engine being compounded, by adding a new Corliss valve HP cylinder in tandem behind the original cylinder. Changes involved the removal of the original LP Meyer valve and additions for operating the Corliss valves – a drag link/return crank connected to a short "outside" shaft for the two eccentrics; and a twin belt pulley for driving a new Proell governor which controlled the trip gear. An original condenser may have been replaced by the new cylinder, but since 1894 at least the engine worked non-condensing. The exposed cylinder lagging gave this engine very much the appearance of a working engine rather than a showpiece, as did the hiding of the flywheel in a metal cupboard. Beyond this, the drive was taken through the engine house wall direct into the mill. A single Lancashire boiler, J.J. Horsfield, Dewsbury No.10626 of 1883, remained in use until around 1982, when it was scrapped, although the engine had stopped working in 1965. However, following restoration work after a takeover of the mill by Skopos Design Ltd. in the early 1980s, the engine was for a time steamed fairly regularly.

159

160

161

162. Edlington, Yorkshire Main Colliery – No.1 Shaft (SK544992)

Photo taken/Reference: 28.4.1979/149-36

Type: Horizontal twin cylinder winder

Maker and Date: Markham & Co. Ltd., Chesterfield. No.4289. 1909

Cylinder dimensions: 42ins (originally) x 7ft. Drop inlet valves; Corliss exhaust valves. (The cylinders were apparently rebored to slightly increased diameters.)

Psi: 120

Service: Winding men and materials to Barnsley seam at depth of 915 yards. Downcast shaft. BCC drum 18ft–34ft diameter. (The shaft could also be used for mine car coal winding. $7^{1}/_{2}$tons per wind.)

The two shafts were sunk by the Staveley Iron and Coal Co. Ltd. in 1909–11, with two similar winding engines being ordered from Markham in 1907. Unusually, the two engines were placed in a single long engine hall, with various turbo-compressors and other power plant between. An impressive sight, they were large and powerful engines, the main difference between them being in the drums: No.1, illustrated, mainly for men and materials, had a bi-cylindro-conical drum, claimed to be the largest drum then remaining in Britain; No.2, the coal winder, had a parallel drum, 24ft in diameter. The valve gear was standard Markham for the period: drop inlets and Corliss exhausts driven via Gooch link motion and wrist plates; and there were lengthy piston tail rods with slippers and tubular covers. The exhaust steam went to the LP system. Housed separately behind the engines were sixteen Lancashire boilers 10ft x 32ft, which had chain grate stokers fitted in 1955. By 1979 five were fired by methane from the mine, the Barnsley seam being a very gassy seam, from which in the 1960s the methane had been piped for use in firing the Manvers Main coke ovens. The original 150 psi working pressure had been reduced. In 1979, electric winders were introduced, but plans to deepen the shafts to the Parkgate seam at 1,109 yards apparently came to nothing, and production ceased in 1985.

163. Elsecar, Elsecar Colliery (SE387000)

Photo taken/Reference: 22.8.1986/519-30
Type: Newcomen atmospheric pumping
Maker and Date: 1795
Cylinder dimensions: 48ins x 5ft. Drop valves
Hp: 11.4 bhp **Spm:** 6 **Psi:** 1½–2½
Service: Mine drainage. Bucket pump (18ins dia x 5ft) raising water from the Barnsley seam in a single lift of 43 yds to underground drainage level. Capacity with working conditions in 1918 as above: 50 gallons/stroke or 300 gallons/minute

The 4th Earl Fitzwilliam's large-scale exploitation of coal on his Wentworth estate began with the opening of the Elsecar New Colliery in 1795, the major expense being the pumping engine, now the only surviving Newcomen engine still on its working site in the U.K. A number of builders were involved, including Longden, Chambers & Newton, of Thorncliffe, who supplied, but apparently did not actually make, the 42 inch cylinder. Various alterations were made subsequently. The Butterley Company provided a larger replacement cylinder in 1801, and the beam was replaced twice – by another wooden one in 1811, and in 1836 by a cast iron beam (with parallel motions at both ends), almost certainly by Graham & Co., Milton Ironworks. Remarkably, the engine remained in use until 1923, when electric pumps were put in, and probably on standby until 1931, when it ran for a Newcomen Society visit. The photograph shows the valve gear: from right to left, the plug rod which operates the levers on the two arbors – steam at the bottom and injection water at the top; the injection valve box, with below it the pipe to the bottom of the cylinder; and the cylinder with one of its four corner supports. The cylinder has a sloping bottom and above this, its intended purpose unknown, a flat false bottom with a rectangular opening just above the steam inlet and a small hole in the centre to accommodate the injection pipe. (For a detailed account of this engine see P.J.M. Southworth, *The Beam Engines of Elsecar, Leawood & Middleton Top*, 2006.)

164. Havercroft with Cold Hiendley, The Monckton Coke & Chemical Co. Ltd., Royston (SE374121)

Photo taken/Reference: 24.10 1982/382-25
Type: Inverted vertical single cylinder
Maker and Date: Marshall Sons & Co. Ltd., Gainsborough. No.86991. 1933
Cylinder dimensions: 8¾ins x 10ins. Slide valve
Rpm: 210 (des.)
Flywheel diameter: 3ft 6ins
Service: Unknown. Drove workshop machines?

The first coke works here were erected by the Monckton Main Colliery Company in 1879, with operations being transferred to the Monckton Coke & Chemical Company in 1920. At least two small reciprocating steam engines remained on the site in 1982: a Belliss & Morcom C-type compound of 1909, originally ordered for Maltby Main Colliery and used here until around mid 1979 to drive a standby generator; and the engine illustrated. This was of Marshall's standard "vertical short-stroke" bottle frame type, with slide valve and Pickering type governor; in this case although the governor was presumably originally fitted, the engine appears to have been last used without it. There is evidence that there had been a lineshaft above, presumably driven by belt from the flywheel, but nothing more to suggest the engine's purpose. All that is known of its previous history is that on 15th May 1933, Marshall delivered it to William Smith, a "marine store dealer", Talbot Road, Blackpool. It was said to have been disused at the Monckton coke works since the Second World War.

162

163

165. Huddersfield, John Brooke & Sons Ltd., woollen & worsted manufacturer, Armitage Bridge Mills (SE135135)

Photo taken/Reference: 24.8.1984/464-35
Type: Francis type water turbine
Maker and Date: W. Günther & Sons, Central Works, Oldham. 1905
Hp: 75 **Rpm:** 162
Service: Electricity generation

Armitage Bridge Mills were started as a woollen mill by the Brooke family around 1816–17 and steadily expanded, using water and steam power. Their considerable final extent may be gauged from the front cover illustration of Colum and Giles, *Yorkshire Textile Mills... 1770–1930*, 1992. The two Günther turbines were presumably installed here new in 1905. They were of the horizontal shaft mixed flow Francis type, suitable for a low head. The 75 hp alternators were chain-driven and at full load ran at 750 rpm. The turbines were each mounted on a flat bed, with water supplied by horizontal pipes. The power was controlled by varying the pivoted guide blades, which were linked to the governor, hand-adjustable by the wheel. Also visible in the foreground is the suction pipe, which takes advantage of the fall to the tail water level. In 1984, the turbines were still in regular use, unlike the Belliss & Morcom steam engine, the top of which is just visible. The mills are now in multiple use, but the turbines (overhauled) have since 1985 been exporting power to the grid.

166. Huddersfield, Parkwood Mills Co. Ltd., woollen manufacturers, Longwood (SE104168)

Photo taken/Reference: 6.9.1984/470-7
Type: Inverted vertical compound (enclosed)
Maker and Date: W.H. Allen Sons & Co. Ltd., Bedford. No. R1/82241/2. 1939
Cylinder dimensions: 14ins/22ins x 10ins. Piston valves
Hp: 365 bhp **Rpm:** 428 **Psi:** 150
Flywheel diameter: 4ft 6ins
Service: Electricity generation. Driving 250 kW General Electric Co. Ltd., Witton alternator

This was a large 19th-century room and power mill, but by 1984 at least, Parkwood Mills Co. Ltd. seemed to be the main user. It was notable for its remarkable 1877 chimney – square for three-quarters of its height, then octagonal. Supplied new to Parkwood Mill, and direct-coupled to an alternator, the engine stopped working around the mid 1970s when it became impossible to make use of the exhaust steam (at 10 psi back pressure). Unlike W.H. Allen's smaller two-crank compound engines, in which two central valves were driven from a single eccentric, those of 220 bhp and above have valves operated by separate eccentrics – one at each end of the engine. In around 2000-2001, the mill's use for textile manufacture came to an end, and by 2005 its conversion for residential use had been approved, although with the chimney being retained.

167. Huddersfield, James Shires & Sons Ltd., woollen & worsted manufacturers, George Street Mills, Milnsbridge (SE117160)

Photo taken/Reference: 11.8.1989/635-36
Type: Inverted vertical twin cylinder (enclosed)
Maker and Date: Ashworth & Parker Ltd., Bury. No. 2044. 1953
Cylinder dimensions: 12ins x 7ins. Piston valves
Hp: 150 bhp **Rpm:** 500 **Psi:** 150
Flywheel diameter: 3ft 6ins
Service: Electricity generation. Direct-coupled to Crompton Parkinson Ltd. alternator. 120 kW

This mill became notable for the late installation in 1980 and 1983 of two reconditioned Belliss & Morcom compound engines (of 1939 and 1907 respectively). Exhausting to process and heating, these could be run economically during the winter months. This success prompted the installation in 1988 of a third engine, the Ashworth & Parker illustrated, which had spent its previous working life at a hosiery dyers, The Springfield Hosiery Finishing Co. Ltd. – later Springfield (Dyers & Finishers) Ltd., Bulwell, Nottingham. However, perhaps not surprisingly, the steam consumption of this engine (14,000lbs/hour) was found to be too high, and it was hardly used. In 1997, the older buildings, south of George Street, in which the engines were located, were sold off for development and then destroyed by fire while being demolished. No definite information as to the fate of the three engines has come to hand, but it seems unlikely that any of them have survived.

165

166

168. Leeds, Central Electricity Generating Board, Whitehall Road Power Station (SE296333)

Photo taken/Reference: 30.4.1979/153-24
Type: Inverted vertical triple expansion (enclosed)
Maker and Date: Belliss & Morcom Ltd., Birmingham. No.1388. 1901
Cylinder dimensions: 19ins/28ins/43ins x 16ins. Piston valves
Hp: 750 bhp **Rpm:** 320 **Psi:** 200
Flywheel diameter: Small flywheel for barring
Service: Pumping water from the River Aire through the turbo-alternator condensers

Around 1900, five Belliss & Morcom triple expansion engines were built to drive alternators for the City of Leeds Electric Lighting Works: Nos. 1286–7 & 1388 (1,000 hp each); and No. 1500 & No. 2242 (2,000 hp each). While the larger engines were removed many years ago, two of the smaller engines (Nos. 1287 and 1388) were retained for standby use until as late as 1968, having in 1915 been converted to provide condenser water for the adjacent turbo-alternators (No. 1286 had been sold on in 1916). The conversion involved their removal from what became the original turbine house to a new coldwater pump house. The alternators were scrapped and direct-driven Rees Roturbo centrifugal pumps fitted, and although the power required was lower, the steam pressure was raised from 150 psi to 200 psi with a higher temperature of up to 500°F. They had surface condensers which were provided with a steam operated air ejector. The engines were last run around 1973–4, by which time the Station had become a CEGB training school. In 1992, with the remaining buildings due to be demolished, the engines were removed to the Markham Grange Steam Museum.

168

169. Leeds, T. Horn, whiting & putty manufacturer, Thwaite Mills, Hunslet (SE328313)

Photo taken/Reference: 21.7.1990/654-17
Type: Steam derrick
Maker and Date: Butters Brothers, Glasgow. No. 2174? 1946
Cylinder dimensions: c.5ins x 8ins (vertical twin). Slide valves with reversing links
Psi: 100
Service: Loading and unloading canal boats in the Mills basin

The main mill buildings are the result of a complete rebuild of an earlier mill by Hewes and Wren for the Aire & Calder Navigation Co. in 1823–5. In 1872, the Horn family took over the lease, manufacturing processed flint and china stone until 1915, and whiting and putty – the latter a mixture of whiting and linseed oil, started in 1923. The derrick was a 3 ton all steel derrick crane installed in 1947 as a replacement of a similar Butters Bros. Scotch derrick put in c.1900. A single twin cylinder engine provided the power for: (1) Hoisting, via double reduction gearing; (2) Derricking (raising and lowering jib) via double reduction plus step up gearing (3) Slewing via double reduction to shaft from which either of two facing bevels drove to a vertical shaft, further reduction geared to a pinion engaging the gear ring at the base. Sadly, on 29th January 1975, a flood washed away half the weir across the River Aire, depriving the two wheels (the main power source – apparently installed in 1876 by Pollit & Wigzell!) of their water supply; although the small putty mills driven by a 25 hp diesel engine continued for a while, at the end of 1976 all operations at this historic site came to an end. The derrick was put into a museum store in Burslem, but 1990 saw a very welcome re-instatement with a replacement boiler as part of the Thwaite Mills preservation project.

170. Linton, Linton Mill (ex Lowcock & Sons Ltd., cotton goods manufacturer) (SE002633)

Photo taken/Reference: 19.4.1981/260-14
Type: Uniflow
Maker and Date: Newton, Bean & Mitchell, Bradford. 1921
Cylinder dimensions: 20ins x 27ins. Piston drop valves
Hp: 300 ihp **Rpm:** 140 **Psi:** 160
Flywheel diameter: 11ft
Service: Driving weaving shed by 10 grooved rope pulley (6ft dia) and 100 kW generator from belt pulley (9ft dia)

Linton Mill was taken over by the Lowcock family in 1911, and after a fire in the following year was completely rebuilt with a new steam engine being installed in 1921. From then until its closure in 1959 it was run as a cotton cloth manufacturing mill by Frank Lowcock. In combination with water turbines and diesel engines, the steam engine also formed part of the private generating plant (belonging to Linton Mill Estates Ltd.) supplying power to North Craven Hydro Electric Co. Ltd. for villages in the area. One of only eleven uniflow engines built by Newton, Bean & Mitchell, it had positively driven inlet valves driven by eccentrics on the bevel gear driven sideshaft and controlled by Eastwood's isochronous shaft governor. Auxiliary horizontal exhaust valves below and at right angles to the cylinder were operated by separate eccentrics when there was insufficient vacuum. The main drive was via a rope driven lineshaft running externally above the River Wharfe along the length of the mill, with a series of bevel gear drives to the weaving shed shafting. The engine was probably stopped in the early 1950s, but even after production was stopped around 1958–9 remained hidden from view until after the death of Mr. Lowcock. The mill was then sold for development, and in 1984 the engine went to Bradford Industrial Museum – a sad end for a fascinating site.

171. Moss, Moss Brick Works (SE586143)

Photo taken/Reference: 18.8.1973/4-26
Type: Horizontal single cylinder
Maker and Date: Unknown
Cylinder dimensions: c.16ins x 3ft. Slide valve; Meyer expansion valve
Flywheel diameter: 9ft
Service: Drove generator, grinding pan (by Wootton Bros, Coalville), and possibly other brickmaking machinery

Unfortunately, no information on the history of this site or its engine has come to hand. By 1973, the works was derelict, although there were still some kilns and chimneys, and the engine, open to the elements, was disintegrating. The only possibly identifying mark on the engine was the figure 564 cast onto the slide valve cover. Part supported by an impressive masonry wall, it was of a fairly simple design with 4-bar crosshead guide and disc crank. The wording on the base of the missing governor shows it to have been a 3 inch Pickering type, and the cylinder rear end cover apparently provided for a tail rod, also absent. The flywheel was grooved for 7 ropes (remains of only 4 still present) driving to an 11-grooved second motion shaft pulley behind the engine, from which 3 ropes came forward to the small generator pulley just visible in the photograph. The second motion shaft was also connected by belt to a third shaft from which a fourth shaft to the edge runners in the grinding pan was driven by worm gearing. The engine was scrapped in 1977.

169

170

171

172. Orgreave, Orgreave Colliery – No. 1 Shaft (SK424871)

Photo taken/Reference: 20.2.1982/315-12

Type: Horizontal twin cylinder winder

Maker and Date: Markham & Co. Ltd., Chesterfield. No. 5931. 1926

Cylinder dimensions: 39ins x 5ft 6ins. Drop inlet valves; piston exhaust valves

Hp: c.1300 **Rpm:** 69 (max) **Psi:** 95

Service: Mainly coal winding from depth 1200ft. Downcast shaft. Parallel drum 18ft diameter. 5 tons per wind

Orgreave Colliery, dating from as early as 1851, was in the hands of the United Steel Companies Ltd. by the time this Markham winder, a relatively late example, was installed in 1927. As far as can be judged from recent survivals, this was not typical Markham. The drop admission valves and large horizontal piston valve exhausts operated by a single wrist plate on each side were not too uncommon, particularly on the later Markhams. But this engine was different in that it had short "outside" eccentric shafts driven by drag links and return cranks. A rocking shaft (driven via Allan link motion rather than the Gooch motion more usual with Markham) passed over the cylinder to a lever and rod connection to the wrist plate. A Hartnell type governor controlled the inlet valve trip gear. There were no tail rods. It raised men as well as coal, for which a very rapid winding rate of 65ft/sec was possible. Coal production ceased on 12th October 1981, and, after salvaging work which continued into 1982, the engine was scrapped in 1983/4.

173. Otley, William Barker & Sons Ltd., tanners & curriers, Cross Green (SE207456)

Photo taken/Reference: 1.5.1979/155-37

Type: Horizontal single cylinder

Maker and Date: Marsden's Engines Ltd., Union Foundry, Heckmondwike. No. B214. 1907

Cylinder dimensions: 17$^{5}/_{8}$ins x 3ft. Corliss valves

Hp: 100 ihp **Rpm:** 96 **Psi:** 50

Flywheel diameter: 10ft

Service: Driving tannery machinery and 18 kW dynamo

This was the last Corliss valve engine, and indeed the last mill engine of any size at all, to remain at work in the U.K., and this wonderful little works was the destination of many a pilgrimage in its final years. Rated according to the maker's original specification to develop 100 ihp at 90 rpm and with a steam pressure of 70–80 psi, from 1973, with a replacement oil-fired boiler, working pressure 50 psi, the engine was running at 96 rpm. It was non-condensing, exhausting to process at c.8 psi. Heavily built, with double slipper guide, it had separate eccentrics for steam and exhaust valves – only the former with a wrist plate. The governor, Watt type but with a centre ball, controlling the trip gear, was mounted on a fluted column. It was tightly fitted into its own engine room inside the main building, within which were the various machines – grinding mills, revolving drums and cutters – driven via lineshaft and a profusion of belts. Drive was from the flywheel, grooved for five 1$^{5}/_{8}$ inch diameter ropes (three on when seen). It showed little sign of care, yet apart from a rebore (the cylinder was originally 17 ins diameter) as far as known seems to have given 80 years' reliable service. It finally stopped on 6th July 1988, with the works running down to complete closure a few weeks later, and was then removed for preservation.

174. Rotherham, British Steel Corporation, Park Gate Works (SK440950)

Photo taken/Reference: 25.3.1977/60-36

Type: Horizontal single cylinder blowing engine

Maker and Date: Gutehoffnungshütte, Oberhausen. 1906

Cylinder dimensions: c.30ins x 3ft 6ins. Piston valve

Rpm: 60 **Psi:** 150

Flywheel diameter: 20ft

Service: Blast furnace blowing engine. Blowing cylinder c.5ft 6ins diameter – discharging 6,000 cubic ft/min at 7 psi

All was derelict by 1977, but six engines were still to be seen: the Henry Berry pumping engine and the Belliss three cylinder compound (see *Stationary Steam Engines of Great Britain*, Volume 1, Nos. 154–155); the Markham pumping engine (see *The Steam Engine in Industry – 2. Mining and the Metal Trades*); two Belliss & Morcom compressors; and the engine illustrated here. Originally a gas engine, presumably installed here new by the Park Gate Iron & Steel Co. Ltd., it had been converted to steam by the Park Gate company – possibly in 1914. The crankshaft had incised on its end: VICKERS LIMITED 1914 25305 NICKEL CHROME. It may have remained in use until the last two furnaces were blown out in 1971–3. By 1975 the works had closed completely and in 1977/8 the site was cleared. Massively built, and of immense length, it had a double web counterbalanced crank, single slipper crosshead guide, single slipper support between steam and blowing cylinders, and finally a tailrod with another slipper. The piston valve spindle, also with a tail, was operated by an eccentric rod and intermediate lever taking the motion to the required higher level. On the other side, next to the flywheel, another eccentric rod operated the blowing cylinder Corliss suction valves. The governor was disconnected but may have been linked to a throttle valve.

172

173

175. Scissett, G.H. Norton & Co. Ltd., pile fabric manufacturers, Nortonthorpe Mills (SE247101)

Photo taken/Reference: 8.8.1983/422-36
Type: Horizontal tandem compound
Maker and Date: Pollit & Wigzell & Co., Bank Foundry, Sowerby Bridge. 1886
Cylinder dimensions: 13¼ins/25¼ins x 5ft. Corliss valves (HP); slide valve (LP)
Hp: 250 **Rpm:** 55 **Psi:** 85
Flywheel diameter: 14ft
Service: Standby drive for generator

A relatively large number of Pollit & Wigzell engines continued to work into the 1970s – a tribute to their high quality – and this is the earliest known to survive. Built when installed here as a slide valve engine, this was their 3-rod type tandem compound, with two LP piston rods passing either side of the HP cylinder, and known by Pollit's as their 'patent engine'. The cylinders were 15¾ins and 28½ins in diameter, the HP fitted with an additional (governor-controlled) expansion valve. Around 1929-31, two new higher pressure Lancashire boilers by J. & J. Horsfield of Dewsbury were installed, and the engine was rebuilt with smaller cylinders to take steam at up to 120 psi. The HP cylinder had separate eccentric rods and wrist plates for the admission and exhaust valves , and there was a third eccentric for the LP slide valve. The Pollit & Wigzell patent Corliss valve trip gear was controlled by a Whitehead governor. Drive was via two 8ft diameter gear wheels in a wooden cupboard, but by the end of its life, as a standby from c.1970, the mill drive had been disconnected, and the engine was driving a generator. After 1984, when the original company went into receivership, parts of the mill remained in multiple occupation, and in the 1990s the engine, having undergone restoration, could be steamed.

176. Sheffield, Abbeydale Works (ex Tyzack, Sons & Turner, steel refiners and edge tool manufacturers) (SK325820)

Photo taken/Reference: 25.8.1984/465-30
Type: Horizontal single cylinder
Maker and Date: Davy Brothers, Sheffield. 1855
Cylinder dimensions: 15ins x 2ft. Slide valve
Hp: 15? **Psi:** 60?
Flywheel diameter: 11ft 2ins
Service: Drove grindstones for sharpening scythe blades

Activity on this site dates back at least to 1676, but the existing buildings date from c.1785–1842. William Tyzack & Sons took over in 1849 and installed the engine in 1855 – to provide supplementary power for the grinding shop of 1817. It was direct-coupled to a lineshaft (also driven by gearing from a waterwheel) carrying three wooden drums from which six grindstones were driven by belts. There were also drives to three additional grindstones via bevel gearing and to two glazers. It had a slide valve on top of the cylinder, driven by a single eccentric rod via rocking shaft and levers, and a bracketed bush guide for the valve spindle. Other features were the counterbalanced single web crank, a flywheel with rather spindly curved spokes, and directly below the crankshaft an eccentric driven vertical boiler feed pump. It seems doubtful if the engine ever saw much use over the long period during which it remained here. When Tyzack, Sons & Turner abandoned the works in 1933 it came into the hands of the City of Sheffield and there followed a thirty-year period of neglect. However, it is now a museum, the Abbeydale Industrial Hamlet, and the engine has been restored.

175

176

177. Slaithwaite, Elon Crowther & Sons Ltd., woollen manufacturers, Upper Mills (SE076138)

Photo taken/Reference: 23.3.1977/58-8
Type: Horizontal tandem compound
Maker and Date: Pollit & Wigzell Ltd., Sowerby Bridge. 1910
Cylinder dimensions: 14½ins/27ins x 33ins. Corliss valves (HP); slide valve (LP)
Hp: 250 ihp **Rpm:** 83–84 **Psi:** 150
Flywheel diameter: 9ft 10½ins
Service: Driving spinning and weaving machinery, both mechanically and via an alternator providing electric power for motor drives

This 3-rod Pollit tandem had been built for Edwards & Rawson Ltd., Canal Mills, Sowerby Bridge, being designed for 60–120 psi steam pressure and 90 rpm, and fitted with a plain 10ft diameter flywheel. In 1924, with the flywheel grooved to take five ropes of 1⅝ins diameter, it was moved to Upper Mills, which had just been taken over by the newly established Elon Crowther company. At the same time, presumably, the steam pressure was raised to 160 psi. Another alteration, in 1931, was the fitting of the Lumb governor and regulator. The drive to the mill was taken via reduction gearing (3ft 6ins:9ft) to a second motion shaft, from which it was transmitted by bevel gearing to a vertical shaft in an exterior tower. The shaft visible in the photograph running above the engine, driven by five ropes from the engine flywheel, was connected by 18 ropes to an Electric Construction Co. 200 kVA alternator. Electric power could also be provided by a Belliss & Morcom generator set (150 kW) and a water turbine (40 kW). Also visible in the photograph above the jet condenser/air pump is an oil separator in the exhaust steam pipe. The engine ran practically day and night, 5½ days a week, until 1966, and then remained on standby for electricity generation only, until 1981, being removed for preservation in 1987. It is now at Masson Mill, Derbyshire, but not yet re-erected.

178. Thurcroft, Thurcroft Colliery (SK499897)

Photo taken/Reference: 11.3.1973/4572
Type: Horizontal cross compound compressor
Maker and Date: Walker Brothers (Wigan) Ltd., Pagefield Ironworks. No. 37919. 1936
Cylinder dimensions: 15½ins/28ins x 3ft. Drop valves
Hp: 659 ihp (des.) **Rpm:** 130 (des.) **Psi:** 160
Service: Colliery air compressor

This Walker Brothers two-stage air compressor, a relatively late example, was still on standby in 1973, but scrapped just over two years later. The maker's records show it to have been ordered on 26th August 1936 by United Steel Companies Ltd. (Rothervale Collieries Branch) for delivery to Orgreave Colliery. Orgreave and Thurcroft were both in the hands of the Rothervale company – confirmation that the engine was actually erected at Orgreave and then transferred has not come to hand. The design specification provided for steam at 190 psi and 500–550°F, working conditions as above, and a capacity of 4,000 cubic feet of free air/minute at 85 psi gauge. Steam was to exhaust at 5 psi (there was no condenser). Drop valves were operated by eccentric rods from the sideshafts, with the governor linked to the air pressure controlling the HP trip gear. Air cylinders, in tandem, were 31½ ins (1st stage) and 19½ ins diameter. Giving increased efficiency, there was a cylindrical intermediate air cooler (with water cooling) between the two stages.

179. Thurnscoe, Hickleton Colliery – No.3 Shaft (SE465053)

Photo taken/Reference: 24.8.1986/523-11
Type: Horizontal twin cylinder winder
Maker and Date: Markham & Co. Ltd., Chesterfield. 1921
Cylinder dimensions: 38ins x 7ft. Drop valves
Hp: 3,090 ihp **Rpm:** 30 **Psi:** 140
Service: Winding mostly men and materials (in the 1970s) from the Parkgate seam (winding depth 2,390ft). Double-decked winding cage (capacity 56 men). Upcast shaft. BCC drum 16–28ft diameter

Initial sinkings of Nos. 1 and 2 Shafts occurred in 1892–4 and of No.3 in 1921–3. (The Nos.1 and 2 winders had been electrified in 1961-2.) The No.3 engine, ordered by the Hickleton Main Colliery Co. Ltd. in 1920, was one of Markham's later standard type. The drop inlet and exhaust valves were driven via Gooch link reversing gear and a round wrist plate on each side, and there were the usual piston tail rods in tubular covers. The original cylinder diameter specified in Markham's order book was 35 inches – was this changed or were the cylinders replacements? It appears from the photograph that the bi-cylindro-conical drum had at most only three parallel turns at the smallest (16ft) diameter. In 1976 the brakes were modernised, and (probably at the same time) the Thornewill & Warham controller replaced. The miners' strike brought production to an end in 1984, but the No.3 Shaft was retained, in part for access and ventilation of the adjoining Goldthorpe Colliery. By then an electric replacement was on site, but the life of the steam winder (converted to compressed air in 1986) was prolonged by continuing uncertainty about the future. Development work at Hickleton having by 1988 proved abortive, most of its surface buildings were then swept away, but two winders and headgears survived until Goldthorpe's closure in 1994.

177

178

179

180. Walton, Walton Colliery (SE360182)

Photo taken/Reference: 20.4.1981/263-4
Type: Inverted vertical compound compressor (enclosed)
Maker and Date: Belliss & Morcom Ltd., Birmingham. No. 1445. 1922
Cylinder dimensions: 23ins/33ins x 17ins. Piston valves
Hp: 715 ihp **Rpm:** 250 **Psi:** 170 (des.)
Service: Supplied compressed air – acting as a standby

The two-stage compressor, Belliss & Morcom type E21, became the standby to an electrically-driven Ingersoll-Rand compressor, but appeared to be disused by 1977. The air cylinders, 36 ins diameter (1st stage) and 21 ins (2nd stage), were able to provide 4,000 cubic feet/minute at 80 psi pressure. They were in tandem, in this case below the steam cylinders (in contrast to those of the compressor at GEC Machines, Rugby – see No. 146). Renamed in 1957, Walton was begun as Sharlston West Colliery by the New Sharlston Collieries Co. Ltd., with the sinking of the first shaft being started in 1890. (For the Fowler engine dating from this period see *Stationary Steam Engines of Great Britain*, Volume 1, No. 212.) But the great expansion came in the early 1920s. The Robey winding engine on No.2 Shaft, the electric winder on No.3 Shaft, and the Power House all date from 1923, the latter still containing the compressors, steam fan engine and two turbo-alternators: Belliss & Morcom mixed-pressure turbines Nos. 242 of c.1926 and 693 of 1941. These drove 1875 kVA and 2500 kVA alternators, the former still on standby in 1973, the latter then out of use.

181. Walton, Walton Colliery (SE360182)

Photo taken/Reference: 26.3.1977/62-36
Type: Inverted vertical triple expansion (enclosed)
Maker and Date: Belliss & Morcom Ltd., Birmingham. No. 6984. 1923
Cylinder dimensions: 16½ins/23ins/33ins x 16ins. Piston valves
Hp: 500 bhp (des.) **Rpm:** 240 (des.) **Psi:** 150/175 (des.)
Service: Direct-coupled to double inlet Davidson "Sirocco" fan. To the right is Belliss & Morcom No. 7745, a C-type compound of 1924, direct-coupled to a 100 kW Crompton alternator

This was a Belliss & Morcom Type T10S ordered by the New Sharlston Collieries Co. Ltd. in 1922, located at one end of the Power House and the standby engine for the fan. The fan house was situated between this and the No.3 Shaft (electric) winding engine house, which also contained the main electric fan motor. The steam winding, compressor and fan engines exhausted to accumulators supplying the mixed-pressure turbines. Walton Colliery was notable for the variety of steam plant still on site and in use, and for its distinguished architecture. All more or less intact when production ceased on 3rd December 1979, this was arguably the finest remaining example of an early 20th-century coal mine in the U.K. Some recognition of this was that Wakefield Metropolitan District Council commissioned a report on the various possibilities for its preservation – tragically nothing came of this, and in 1982 all was swept away.

182. Yeadon, James Ives & Co. Ltd., woollen manufacturers, Manor Mill (SE205412)

Photo taken/Reference: 30.4.1979/154-20
Type: Horizontal tandem compound
Maker and Date: Pollit & Wigzell Ltd., Sowerby Bridge. 1920
Cylinder dimensions: 18ins/34ins x 4ft. Corliss valves
Hp: 400–600 ihp (des.) **Rpm:** 75 **Psi:** 100–160 (des.)
Flywheel diameter: 15ft 6ins
Service: Drove mill and 210 kVA English Electric alternator

The woollen manufacturers, James Ives & Co. had two mills in Yeadon: Leafield Mill, which they took over around 1884; and Manor Mill, leased during the period 1865–84 and then purchased in 1911. Pollit & Wigzell engines, well-known for their high quality, appear to have been favoured for the power plant. Three tandem compounds were supplied: for Leafield Mills, a 250 hp slide valve engine in 1885, and a 350 hp Corliss and slide valve engine in 1903; and for Manor Mill, the engine illustrated here. All appear to have given good service – the 1903 engine (illustrated in *Stationary Steam Engines of Great Britain*, Volume 1 – see No. 221) lasted into the early 1970s, and the Manor Mill engine continued in use until the last week of July 1976. Unlike the other three tandem Pollits illustrated in this volume, both cylinders had Corliss valves, and slightly unusually this was not of the patent 3-rod type – it had a single piston rod and the cylinders were separated. A single eccentric rod operated both HP and LP admission valves via two linked wrist plates, and the HP and LP exhausts each had a separate eccentric and wrist plate. A Lumb governor and regulator controlled the HP trip gear. The mill was driven by up to 14 ropes 2ins diameter from the flywheel in a separate room, but by 1973 only 6 ropes were in use, the steam pressure was down to 100 psi or less and the engine was probably developing under 300 hp. Unfortunately, this fine engine was scrapped following closure of the mill in 1980.

180

181

182

183. York, The York Waterworks Company, York Waterworks, Acomb Landing (SE583525)

Photo taken/Reference: 27.4.1974/14-10

Type: Horizontal duplex tandem triple-expansion non-rotative pumping

Maker and Date: James Simpson & Co. Ltd., London & Newark. No.2812. 1907

Cylinder dimensions: 16ins/25ins/42ins x 2ft. Semi-rotary slide valves; HP cylinders each with two auxiliary cut-off valves

Hp: 246 ihp **Rpm:** 28 **Psi:** 100

Service: Water supply of York. Low-lift pumps (22½ins plungers, capacity 6¾ m.g.d.) delivered water from the River Ouse into subsiding reservoirs; high-lift pumps (21½ins plungers) pumped filtered water into the district mains. Total head 200ft

The Acomb Landing Station had been started in 1846 with two Cornish engines by the Low Moor Ironworks Co. Installed later were two Hawks, Crawshay beam engines, 1879; a James Simpson horizontal cross compound, 1895; and the Worthington, the last to survive. Even this became standby after the installation of oil engines driving centrifugal pumps – four by Ruston, Hornsby in 1923 and three by Vickers-Petters, 1926, and from 1930 electrically driven pumps. However, it remained workable until at least 1952, when it was steamed for a Newcomen Society visit. By 1974, it was missing valve rods and much of the water end, and even these remains were scrapped within the next two years. (The oil engines, also, were due to be phased out.) The arrangement of the pumps was of interest. The two high-lift pumps, largely intact, were in line with the steam cylinders. No longer present were the two low-lift pumps, below floor level and driven by rocker beams from the high-lift pump tail rods. The low-lift pump tail rods had in turn driven two double-acting air pumps for a surface condenser.

CAERNARVONSHIRE

184. Llandwrog, Glynllifon Estate Sawmill/Workshop (SH454554)

Photo taken/Reference: 10.9.1996/822-2

Type: Horizontal single cylinder

Maker and Date: Probably Thomas & De Winton, Caernarvon. 1854

Cylinder dimensions: 13ins x 27ins. Slide valve

Psi: 40

Flywheel diameter: 10ft

Service: Drove reciprocating saw and stone cutting bench

Much of the sawmill-workshop complex of Lord Newborough's Glynllifon estate, including the engine, remains intact. The Cornish boiler (c.4ft 9ins x 15ft) has a plate: THOMAS & DE WINTON CARNARVON 1854 – a maker and date which almost certainly also apply to the steam engine. Typical of the mid 19th century are the flat bed, and the slide valve in a flat box over the cylinder – operated via a gab link to allow for manual working, and a lever and rocking shaft. Other features are the wood strip cladding of the cylinder, a very prominent Watt governor just in front of the crank, and a horizontal boiler feed pump driven from one end of the crosshead. A reciprocating saw (removed in 1987) was driven by belt from an extension of the main shaft into the next room. Still present is a slate cutting bench which was driven by belt from a pulley (4ft 11ins diameter) next to the flywheel. Disused since 1932, as a result of boiler failure, the engine was restored to working order (with a newly installed vertical boiler) by Fred Dibnah in 1991.

185. Nantlle, Blaen-y-cae Slate Quarry (SH497537)

Photo taken/Reference: 11.10.1981/311-14

Type: Diagonal twin cylinder winch

Maker and Date: John M.Henderson & Co., Aberdeen. c.1910

Cylinder dimensions: c.10ins x 18ins. Slide valves; Stephenson reversing links

Service: Blondin aerial cableway winch

A large number of Blondin cableways were introduced into the Nantlle valley slate quarries, following Henderson's patent of a new system in 1896. The steam winch at Blaen-y-cae, a small quarry worked from around 1830 to the 1930s, appears to be of the standard type developed by Henderson. It has two drums – one wider, parallel drum, 5ft in diameter, and a narrower one. These could be operated independently, each having its own driving pinion (reduction c.4:1), which could be clutched in or out of gear. In the Blondin system, the load carriage is supported on a wire between two masts and is drawn backwards and forwards along it by an endless travelling rope, which is wound four or five times round the narrow drum. The wider drum operates another rope which raises and lowers the hoist pulley beneath the carriage. (Further details may be found in Gwynfor Pierce Jones, "The Slate Quarries of the Nantlle Valley", *Stationary Power* No. 2, 1985.)

183

184

185

186

CARMARTHENSHIRE

186. Kidwelly, Kidwelly Tin Plate Works (SN422079)

Photo taken/Reference: 13.4.1976/45-21
Type: Horizontal tandem compound
Maker and Date: Cole, Marchent & Morley Ltd., Bradford. No. W4692. 1919
Cylinder dimensions: 18ins/32ins x 3ft. Piston drop valves
Hp: 300 **Rpm:** 80 **Psi:** 90
Flywheel diameter: 18ft
Service: Driving cold rolls

One of the much earlier Foden engines driving the hot rolls was illustrated in *Stationary Steam Engines of Great Britain*, Volume 4 (see No. 15). In 1904 the Kidwelly Tin Plate Co. Ltd. had taken over and in 1919 put in a new cold rolling plant driven by the Cole, Marchent & Morley engine. This was typical of this company during its last years, having a mammoth bed and Morley's piston drop valves (patented in 1904). They were driven by eccentric rods from the bevel gear driven layshaft, with the HP trip gear controlled by a Hartnell governor (gone by 1976). The jet condenser and horizontal air pump were in tandem behind the LP cylinder. The 8ft 10ins diameter pulley adjacent to the flywheel was grooved for fifteen 1¾ins diameter ropes, five to each of the three 14ft 6ins diameter mill shaft pulleys (the remains of the mills lay outside in the open). The power output appears modest for an engine of this size, but may be accounted for by the relatively low steam pressure used – possibly the boilers had been downrated. All was stopped in July 1941 and in March 1946 the works was dismantled under the Tinplate Redundancy Scheme – but the engine apparently remained in quite reasonable condition for many years. However, some deterioration had occurred by 1976, and when seen again two years later it had suffered severely from the depredations of metal thieves. The works has now become the Kidwelly Industrial Museum.

DENBIGHSHIRE

187. Erddig, Erddig Estate Sawmill (SJ326482)

Photo taken/Reference: 23.7.1977/75-29
Type: Vertical single cylinder
Maker and Date: Probably 1861
Cylinder dimensions: c.6ins x 16ins. Slide valve
Flywheel diameter: 5ft 2ins
Service: Drove sawbench; mortar mill; corn mill; and workshop machinery

The Erddig estate was owned by the Yorke family from 1733 to 1973, when it was passed to the National Trust. The engine, another now rare example of a true vertical (crank overhead), was installed in the early 1860s (possibly 1861), when the estate was in the hands of Simon Yorke III (1811–94). A double A-frame supported the main shaft and there was a pitchfork connecting rod to allow for the bush guided piston rod extension. The steam exhausted to atmosphere via a feedwater heater. Beyond the flywheel, on the other side of the wall, are the belt pulleys for driving sawbench and mortar mill, and then further lineshafting hung from the ceiling. The sawbench was by T. Robinson & Son Ltd., Rochdale (presumably 1880 or later); the mortar mill, a small pan with two edge runners standing outside in the open, was apparently installed in 1900, and used for mixing ashes, lime and sand to make a black mortar for estate use. The small Cornish boiler (still in situ)) was superseded by a second-hand water tube boiler in 1935. According to Merlin Waterson (*Country Life*, 5 Oct 1978) the use of the sawmill was much increased in the 1940s and 1950s. A disagreement with the National Coal Board resulting from damage caused by underground mining led to Simon Yorke IV (1903–66) insisting that only wood be used in the house for fuel, the sawmill being used to cut up estate wood. The boiler, however, did use coal. So far it has been impossible to discover when the engine went out of use – there is an oil engine which could have been used in its place.

188. Trevor, Roberts & Maginnis Ltd., Brick Works (SJ265425)

Photo taken/Reference: 10.7.1979/162-10
Type: Horizontal single cylinder
Maker and Date: Pollit & Wigzell Ltd., Sowerby Bridge. 1923
Cylinder dimensions: 17ins x 3ft. Corliss valves
Hp: 150–200 **Rpm:** 80 **Psi:** 140 **Flywheel diameter:** 10ft
Service: Driving brickmaking machinery by rope drives from the flywheel to two lineshafts

This brickworks, known as the Australia yard, dated from c.1860 and was taken over by Roberts & Maginnis probably in the 1880s; by 1914, taking advantage of good fireclay and ganister deposits, the production of building bricks had been abandoned in favour of silica bricks, used as refractory linings in the steel industry. The engine, now preserved at Ironbridge, was the only single cylinder Pollit (of 16 built) known to have survived into recent years. There has been a recent suggestion, not confirmed, that it was first installed (although never worked) in a northern cotton mill, and then purchased second-hand for installation here a few years later – a price of £750 has been quoted (see Andrew Connolly, *Life in the Victorian Brickyards of Flintshire and Denbighshire*, 2003). Perhaps in appreciation of its quality, and unlike some brickworks engines, it seems to have been well looked after. It had apparently given trouble-free service, although in 1974 the cylinder was given a rebore and a new piston, and new ropes were fitted. Other than that it seems to have been unchanged from new. Pollit & Wigzell's usual Corliss valve trip gear was controlled by a Whitehead governor and, working without a condenser, the exhaust steam was used for brick drying. The drives were taken by 10 ropes from the flywheel, split 6 and 4 to separate lineshafts, driving two edge runner mills, and various crushers, fans and elevators. It continued working until production ceased in mid January 1980, the market having disappeared with the closure of the Shotton steel works, where ROMAG bricks had been used for lining the open hearth furnaces. The engine was then purchased (for £3,100) to drive a rolling mill in the Blists Hill Museum.

FLINTSHIRE

189. Rhydymwyn, Ruby Brick & Tile Co. Ltd., Ruby Brickworks (SJ205677)

Photo taken/Reference: 1.11.1974/22-33
Type: Horizontal cross compound
Maker and Date: Robey & Co. Ltd., Lincoln. Nos. 22825–6. 1903
Cylinder dimensions: 13½ins/21½ins x 30ins. Drop valves; slide exhaust valves
Rpm: 90 (des.) **Flywheel diameter:** 11ft
Service: Driving mills and wire-cut brickmaking machinery

This works, started in 1893, was by 1898 being operated by the Ruby Brick & Tile Co. Ltd., which from 1902 was owned by Hartley & Partners Ltd. Initially making high-quality facing bricks, by the 1930s it had switched to common wire-cut bricks. This was the cross compound version of Robey's Class E engine, in this case with mammoth beds. Richardson & Rowlands Patent Cut-Off Gear was fitted, and the governor, standing on the HP cylinder, was driven in the usual manner by gearing from the sideshaft. Drive to the lineshaft was by belt from the flywheel (face 30 inches). Unfortunately, little is known for certain about the history of this engine. It was non-condensing here, but Robey's records show that it was originally built with a jet condenser behind the LP cylinder (which does indeed show signs of having had a tailrod). This suggests that it was second-hand. It was actually initially ordered from Robeys by the Electric Construction Co. Ltd., Wolverhampton, although most likely not for their own use. It has been stated that the engine stopped working in 1936, but this has not been confirmed. With demand falling and antiquated machinery, this small works struggled on until 1965. The buildings remained, mouldering and increasingly ruinous until demolition started in 2003, but the engine had been removed for preservation in 1977. Two years later it was seen in pieces at Grosvenor, Chater's Abbey Paper Mills, Greenfield.

GLAMORGANSHIRE

190. Mountain Ash, Penrikyber Colliery – No.3 Shaft, Penrhiwceiber (ST062969)

Photo taken/Reference: 14.4.1976/46-11
Type: Horizontal twin cylinder winder
Maker and Date: Robey & Co. Ltd., Lincoln. Nos. 30082–3. 1911
Cylinder dimensions: 17ins x 33ins. Drop valves
Service: Coal winding (until 1949). Diabolo drum 7–10ft dia

The initial sinking at Penrikyber occurred in 1872–8, but the No.3 Shaft was not opened until 1912. The engine was delivered by Robeys to the Penrikyber Navigation Colliery Co. Ltd. on 16th February 1911. It was their Class E coupled winder with the usual layshaft driven drop valves and reversing gear, still with its Robey auxiliaries: brake engine and overwind/overspeed prevention gear. The diabolo drum, however, was slightly unusual. When the No. 3 shaft ceased coal winding in 1949, it was retained for pumping. The Robey winding engine also remained in occasional use, although latterly worked by compressed air, perhaps until about 1973. (The steam winders on Nos. 1 and 2 shafts had been replaced by electric winders in the 1950s.) Quite derelict looking when seen, it was taken out shortly afterwards, and after moving between various sites is now being installed at the preserved Foxfield Colliery in Staffordshire. (Penrikyber Colliery ceased production in 1985.)

188

189

191. Trehafod, Lewis Merthyr Colliery – Trevor Shaft (ST039911)

Photo taken/Reference: 28.6.1981/281-8
Type: Horizontal twin cylinder winder
Maker and Date: John Fowler & Co. Ltd., Leeds. No. 6160. 1890
Cylinder dimensions: 42ins x 6ft. Cornish valves; Allan reversing links
Rpm: 11.3 (max. – 1981) **Psi:** c.70 (air)
Service: Winding men and materials. Winding depths 1,078ft and 1,375ft. Differential BCC drum: 10–17ft; 14–22ft

Lewis Merthyr Colliery was initiated in the 1870s by W.T. Lewis (later Lord Merthyr), two of the shafts (Trevor and Bertie) being named after his sons. *Stationary Steam Engines of Great Britain*, Volume 4 included illustrations (see Nos. 65–68) of four of its engines, the last of which to survive being the Coedcae Shaft winder by J.D. Leigh until its scrapping, probably in 1974. Not among these was the Trevor Shaft engine (shaft and winder dating from 1890), which continued in use until production ceased in 1983 and remains on the site today. However, coal winding at Lewis Merthyr had ended in 1958 after a merger with Tymawr Colliery, and shortly after this the Trevor engine was converted to compressed air, the boilers at Lewis Merthyr being scrapped. Presumably at some time before this, the valve gear was altered by Worsley Mesnes, whose plates appear on the dashpots. The Hartnell type governor would have controlled the trip gear, but this has been disconnected. The somewhat unusual differential conical drum apparently made it possible to wind simultaneously to the two different levels. Today, the colliery (or the part remaining) forms the Rhondda Heritage Park, with winders on Trevor and Bertie Shafts – the latter must have been electrified, using the Robey drum of 1930, just after Watkins photographed it in 1959.

MONMOUTHSHIRE

192. Abercarn, British Steel Corporation, Abercarn Tinplate Works (ST214949)

Photo taken/Reference: 16.8.1978/124-3
Type: Uniflow
Maker and Date: John Musgrave & Sons Ltd., Bolton. No. N62/10. 1911
Cylinder dimensions: 23½ins x 33ins. Drop valves
Hp: 400 ihp **Rpm:** 140 **Psi:** 170
Flywheel diameter: 12ft
Service: Driving directly coupled Siemens DC generator. 250 kW

The first practical uniflow engine was built in Germany in 1908 using the Stumpf system, which was adopted in the U.K. by John Musgrave & Sons as early as 1909 and used by them until 1925. This early example was one of two identical engines built in 1911 for Richard Thomas & Co. Ltd., the other being for their Burry Works, Llanelli. In the Stumpf valve gear the double beat drop valves were operated by rollers on an eccentric driven reciprocating rod, the position of which was controlled by a crankshaft governor. Here, there was an underfloor jet condenser provided with an air pump driven by 4 ropes from a pulley on the main engine shaft. The engine may have been stopped when tinplate manufacture ceased in May 1961; it remained on standby until 1981 (being used during the 3-day week period in 1974), but was scrapped in 1985. This engine, an important survival and in excellent condition, should have been preserved – particularly in view of the fate of the similar 1910 engine at the Hafod Works, Swansea (see *Stationary Steam Engines of Great Britain*, Volume 4, No. 64), which continued in use until 1980, was earmarked for preservation, but is now a mouldering ruin.

191

192

193. Bedwellty, Llanover Colliery Pumping Station (SO180008)

Photo taken/Reference: 11.4.1975/45-12
Type: Horizontal tandem compound non-rotative pumping engine
Maker and Date: Hathorn, Davey & Co. Ltd., Leeds. No. 6676. 1914
Cylinder dimensions: 33ins/66ins x 10ft. Piston valve (HP); slide valve (LP)
Psi: 120
Service: Mine drainage. Two single-acting ram pumps (22ins x 10ft stroke) raising water c.500ft to adit 83ft below the surface. Capacity 120,000 gallons/hour

As George Watkins acknowledges, the photograph of this engine in *Stationary Steam Engines of Great Britain*, Volume 4 (see No. 76) is not up to his usual standard, and it seems reasonable to publish another here. Llanover Colliery was sunk by the Bargoed Coal Co. in 1910. This was the derelict and incomplete survivor the larger of two similar engines installed, each one working two pumps at the bottom of the single shaft. The smaller (West) pump had been delivered in 1910. The pumps were operated by a double connecting rod to two quadrants (no longer present) at the top of the well. There was a jet condenser below and a horizontal double-acting air pump driven from one of the quadrants. The speed of the engine was controlled by Davey's differential gear, the remains of which can be seen on the left. The photograph also shows the piston valve above the HP cylinder and higher up to the right a reheater which received the HP exhaust. In 1931–3 the steam pumps were superseded by two electrically-driven Sulzer deepwell centrifugal pumps. At this time the Colliery ceased production, but continued as a pumping station operated by the Tredegar Iron & Coal Co. Ltd. Pumping from the shaft (by then with an even more modern submersible pump) only ceased with the closure of the neighbouring Oakdale Colliery in 1989.

194. Cwm, Marine Colliery (SO188040)

Photo taken/Reference: 7.12.1980/242-9
Type: Inverted vertical twin cylinder geared haulage
Maker and Date: Ebbw Vale Company. 1863
Cylinder dimensions: c.12ins x 20ins. Slide valves; Stephenson reversing links
Service: Endless rope haulage engine

Marine Colliery was sunk 1887–93 by the Ebbw Vale Steel, Iron and Coal Company Ltd., passed through the hands of Partridge, Jones & John Paton Ltd. from 1935 to 1947, and continued to produce coal until 1989. This was one of two geared haulage engines in the same room (the other being a partly buried horizontal twin cylinder engine), apparently derelict and long out of use when first seen in 1973. It has an inscription EVA 1863 on one of the slide valve covers, which has been taken to mean that it was actually built by the Ebbw Vale Company in that year, with the A suggesting that it may have been built at their Abersychan Iron Works. (This company became the Ebbw Vale Steel, Iron and Coal Co. in 1868.) It is thought to have been possibly first used at the Prince of Wales Colliery, Abercarn, also owned by the Ebbw Vale Co. and sunk in 1862–5, before being installed here c.1890. The haulage pulley (c.6ft 3ins diameter) operated an endless rope system, with the rope coiled several times round the pulley and passing down the shaft. It was driven by herringbone reduction gears (2ft and 7ft 6ins diameter). There were counterbalanced single web cranks with twin connecting rods to the crossheads. The engine was moved to the South Wales Miners Museum, Afan Argoed Country Park in 1989.

195/196. Cwm, Marine Colliery (SO188040)

Photo taken/Reference: 7.12.1980/241-17 and 241-27
Type: Horizontal tandem compound non-rotative pumping engine
Maker and Date: Hathorn, Davey & Co., Leeds. No. 5102. 1895
Cylinder dimensions: 36ins/68ins x 10ft. Slide valves; Meyer expansion valves (HP and LP)
Spm: 7 **Psi:** 100?
Service: Mine drainage. Design capacity 50,000 gallons/hour with three lifts totalling 420 yards. Two bucket pumps 12½ins dia, two plunger pumps 13ins, two plunger pumps 13½ins, all 10ft stroke

The pumping engine was installed immediately beneath the winding engine on the downcast shaft. A new LP cylinder and HP piston were fitted in 1910, but it may have had a relatively short life – a stopping date of around 1914–18 has been suggested. It then lingered on, more or less intact, until after the colliery's closure. A rather simpler form of Davey's differential gear, without a pause cylinder, is used here. The engine is controlled by a small subsidiary horizontal engine consisting of a steam cylinder and a cataract cylinder, the piston rod of which is linked by a rocking lever to the main engine valves working over the top of the cylinders. This ensures that the engine works at a constant (though adjustable) speed. The Meyer valves are linked to the main engine crosshead. As shown in the photograph, there had been two HP piston rods (cut off by 1973) extending to the crosshead (only the slippers in place), from which a connecting rod extended across the colliery yard to the shaft, where two opposed quadrants drove the pumps. Of these, only one of the quadrants and its pitwork remained. There was a vertical jet condenser (revealed behind the LP cylinder when the rubble was cleared away prior to the engine's removal c.1992), and according to the builder's records it was supplied with an air pump operated by a separate steam cylinder. However, a *Colliery Guardian* report of 1895 states that the air pump 13ins diameter x 10ft stroke was driven by the LP tailrod!

197. Pontypool, Glyn Pits (ST265999)

Photo taken/Reference: 4.7.1991/672-23
Type: Single cylinder beam pumping
Maker and Date: Neath Abbey Engineering Works. 1845
Cylinder dimensions: 30ins x 6ft. Cornish valves
Rpm: 18–20 **Psi:** c.50
Flywheel diameter: 18ft
Service: Mine drainage, pumping in two stages from a depth of 186 yds to a 380 yds long level, 86 yds from the surface. Capacity 9,000–12,000 gallons/hour

The winding engine at this site, but not the Neath Abbey beam engine, was dealt with in *Stationary Steam Engines in Great Britain*, Volume 4 (see No. 89). The latter was installed for Capel Hanbury Leigh of Pontypool, who sank the pit c.1840–5. It was initially used for winding as well as for pumping (see Marilyn Palmer and Peter Neaverson, "The Steam Engines at Glyn Pits Colliery, Pontypool...", *Industrial Archaeology Review*, Vol. XIII, No.1, 1990). Pumping was by reduction gears (3ft 6ins: 9ft) to a second motion shaft crank operating a horizontal connecting rod to an inverted tee bob carrying a spear rod at the far end and a balance box at the near end. For winding, the crankshaft pinion appears also to have engaged a spur gear on a separate winding shaft carrying two chain barrels. There may have been a second eccentric rod for reversing. All this would have been removed when the vertical winder was installed some time in the period 1856–65. The colliery probably stopped production in 1928, but remained as a pumping station for Hafodyrynys Colliery until the latter's closure in 1966. Possibly the beam engine remained in use until the 1930s, since when it has become increasingly derelict.

194

195

196

PEMBROKESHIRE

198. Pembroke, Ministry of Defence, Llanion Oil Fuel Depot, Pembroke Dock (SM981041)

Photo taken/Reference: 21.8.1978/130-9
Type: Horizontal duplex tandem compound non-rotative pumping
Maker and Date: Worthington-Simpson Ltd., Newark. No.3094. 1928
Cylinder dimensions: 19ins/34ins x 2ft. Semi-rotary slide valves
Rpm: c.50–60 **Psi:** 120?
Service: Pumping fuel oil. Two pumps (11³/₄ins dia x 2ft stroke). Design capacity (at 30 spm): 200 tons/hour with discharge pressure 250 psi

The Admiralty installed a number of Worthington-type pumping engines for warship fuelling installations, but the majority were triple-expansion, and the two installed at Llanion (Worthington-Simpson Nos. 3094–5) were probably the only double expansion compounds. They had worked in the usual way with the twin pumps in tandem and were non-condensing, exhausting to atmosphere. It was said that with both engines working, a maximum of 711 tons/hour of furnace oil could be pumped from Llanion to Llanreath – a distance of approximately 2 miles. When photographed, the engines were still in reasonable condition although stopped since November 1977, but about two years later they were scrapped. The two boilers were of twin furnace "Economic" type by Edwin Danks & Co., Oldbury (Nos. 6492–3 of 1928) and had been oil-fired from the beginning. During the engines' last two years approximately, these had been out of use, and steam at a higher pressure (200 psi) had been supplied from a boiler plant about a quater of a mile away.

ABERDEENSHIRE

199. Aberdeen, James Cordiner & Son Ltd., Victoria Saw Mills (NJ952052)

Photo taken/Reference: 3.9.1975/38-19
Type: Horizontal single cylinder
Maker and Date: Marshall Sons & Co. Ltd., Gainsborough. No. 48835. 1908
Cylinder dimensions: 14¹/₂ins x 3ft. Drop valves
Hp: c.150 **Rpm:** Variable (according to load) **Psi:** c.60 **Flywheel diameter:** 12ft
Service: Driving sawmill for manufacture of fish boxes

This was a relatively early example of Marshall's "L" Class engine, introduced around 1900. It was a quality product for which a minimum steam consumption for a given power output was claimed. As with the smaller engine at Tiptree (see No. 61), there were double beat steam and exhaust valves, the former being controlled by Proell governor and Marshall's patent trip gear, and the exhaust valves driven by cam and roller gear. This was the largest size for which the exhausts were thus operated – cylinders of 16 inches diameter and above normally had the No.2 Gear, in which the exhausts were driven by eccentrics. The two cylindrical fire tube boilers by James Abernethy & Co., Ltd., Aberdeen (one in use at a time) with underfloor furnaces fired with sawdust etc. had a maximum working pressure of 140 psi, but when seen were normally being pressed to only around 60 psi. The drive to the millshaft was by five ropes from the grooved flywheel. Still working when seen, unfortunately the engine was stopped a few months later, and although possibly standby for a few years more, it was scrapped in 1983/4.

200. Cruden, Cruden Bay Brick & Tile Co. Ltd. (NK090369)

Photo taken/Reference: 7.7.1993/729-9
Type: Horizontal single cylinder
Maker and Date: Marshall Sons & Co. Ltd., Gainsborough. No. 38423. 1903
Cylinder dimensions: 18ins (probably) x 3ft. Slide valve; expansion slide valve
Hp: c.150 **Flywheel diameter:** 12ft
Service: Driving brick presses, pipe machine etc.

The Cruden Bay works originally started in 1882, but in 1902 relocated to another site about half a mile away. The Marshall engine was presumably installed new at the same time. It was an engine of Marshall's "K" Class, with which apparently the company replaced its "C" Class around 1900. (For a "C" Class engine see Low Hall Sewage Farm, Walthamstow – No.62.) Both used Hartnell's valve gear and governor, in which an expansion valve was driven by a second eccentric through a governor-controlled linkage. It had a large (nearly 4ft diameter) disc crank and a piston tail rod in a tubular cover. The drive was by 12 inch wide leather belt from the flywheel to a lineshaft, from which belts drove to brick presses and other machinery. The exhaust steam passed through a feedwater heater before use in the drying floors. Modernisation resulted in the stopping of the engine in 1974/5, and it was possibly then that the electric motor drive to the lineshaft was fitted (unfortunately, the engine was never seen in use). However, in 1990, the works was closed, although not cleared away until about ten years later. The single oil-fired Lancashire boiler had been scrapped c.1979, but the engine remained, gathering a thick coating of dust, as shown in the photograph, taken through a hole in the engine room wall. When last seen in 2000, it was in an Aberdeen scrapyard.

197

198

199

201. Skene, Garlogie Mill (NJ782055)

Photo taken/Reference: 7.7.1993/728-19
Type: Single cylinder rotative beam
Maker and Date: Possibly Mitchell & Neilson, Glasgow? 1830s
Cylinder dimensions: c.18ins x 50ins. Slide valve
Hp: c.50? **Flywheel diameter:** 16ft
Service: Drove wool spinning machinery

Garlogie Mill, at first using water power, was built probably in the 1750s for Alexander Hadden & Sons, a woollen manufacturer established c.1748. It was a spinning mill, supplying yarn to the company's main works in Aberdeen, which became the city's largest supplier of carpets (a Boulton, Watt & Co. engine was built for the Aberdeen works in 1805). The Garlogie engine, introduced as a supplementary power source, is understood to date from the 1830s. If the attribution to Mitchell and Neilson of Glasgow (on grounds of design similarities) is correct, it would probably be post 1836/7, when the company started. In 1904, the mill closed and most of the associated village gradually disappeared. Around 1934, the mill itself was largely demolished, but the engine house and contents – a rare and early example of a house-built beam engine (the only one surviving in a textile mill) – have remained. It seems typical of its time, with a fluted column helping to support the single plate cast-iron beam, 16ft 3ins between centres, and with a deep central rib. A lattice work eccentric rod (tied out of the way when photographed) to a rocking lever with two vertical rods passing up each side of the valve chest operated the slide valve spindle from above. The mill drive was presumably via spur gearing from the main shaft which passes through the engine house wall. The photograph clearly shows the evidence of the presence of pigeons, but since then there has been some restoration work by the Grampian Regional Council.

AYRSHIRE

202. Irvine, Matthew Wright & Nephew Ltd., Irvine Sawmills (NS314385)

Photo taken/Reference: 27.8.1979/183-10
Type: Horizontal tandem compound
Maker and Date: Unknown. c.1850–70?
Cylinder dimensions: c.12ins/c.18ins x 28ins. Slide valves
Rpm: 50 **Psi:** 40–60 **Flywheel diameter:** 9ft 4ins
Service: Auxiliary drive for extractor fans (originally drove entire mill)

Unfortunately, nothing seems to be known about the maker of this engine, and the 1850–70 date suggestion is derived from uncertain on site information. However, a somewhat antiquated appearance – its design that of just about the simplest form of compound engine imaginable – does seem consistent with this dating. The HP cylinder was behind the LP, and both had slide valves driven from the one eccentric, with the valve spindle passing from the LP to the HP valve chest. From the LP cylinder, still with its wood strips, distance pieces tapered down to the HP, now completely without lagging. Low on the ground, the engine was bolted to the bedplate sunk into the wooden flooring. Speed was controlled only by a Pickering governor, and out of use, below the floor, was a jet condenser and vertical air pump, driven from one end of the crosshead by a side rod coming forward to an L lever (disconnected by 1975). The mill drive was from a double helical gear (7ft 2ins diameter), located between the flywheel and the wall bearing, to lineshafting just below floor level. Despite its neglected appearance (cosmetic care at least was hardly excessive) the engine was still in working order after a life of probably more than a hundred years – a tribute perhaps to a robust and simple design. In the 1980s it was removed to the Scottish Maritime Museum, also in Irvine.

203. Millport (Great Cumbrae Island), Scottish Gas Board, Millport Gas Works (NS167552)

Photo taken/Reference: 29.8.1979/185-30
Type: Horizontal single cylinder
Maker and Date: The Bryan Donkin Co. Ltd., Chesterfield
Cylinder dimensions: c.6ins x 5ins. Slide valve
Psi: c.75 **Flywheel diameter:** 20½ins (two flywheels)
Service: The No.2 Exhauster. Driving standby exhauster. Capacity 5,000 cubic ft/hour

Some time after all coal gas production in England had ended, five small Scottish horizontal retort works – Golspie, Langholm, Millport, Muirkirk and Newton Stewart – continued to operate into the late 1970s. Millport, with three benches, each of six retorts, was the last to close – on 1st July 1981. This is one of the two exhauster sets, used to pull gas from the retorts and through the hydraulic main and condensers. The No.1's steam engine had been replaced by a 3 hp electric motor in 1959, with a 2 hp petrol engine as standby. The No.2 set was Donkin's basic small horizontal engine direct-coupled to a Beale's rotating vane type exhauster, also by The Bryan Donkin Company. Based on John Beale's patents of 1866 and 1877, and widely used in Scotland from the late 1870s, Donkin had at first been the sole manufacturer, but they were later joined particularly by G. Waller & Son Ltd., Stroud. To maintain the outlet gas pressure (the normal water gauge pressure for the No.1 Exhauster was 12 inches), the throttle valve of the engine was linked to the gas pressure governor – the vertical cylinder about 2ft high, to the right of the engine in the photograph. Unfortunately, the installation date of the engines seems to be unknown, but the Cornish boiler, c.10ft x c.4ft, is by Penman & Co. Ltd., Caledonian Iron Works, Glasgow, 1922. In June 1981, the engine was run for the last time (and for what was said to be the first time for two years), and by mid 1983 the works had been demolished and the engine probably removed for preservation.

200

201

202

CLACKMANNANSHIRE

204. Alloa, Devon Colliery – No.1 Pit, New Sauchie (NS898959)

Photo taken/Reference: 27.8.1975/36-5
Type: Cornish beam pumping
Maker and Date: Neilson & Co., Glasgow. No. 189. 1854
Cylinder dimensions: (72ins x 9ft)
Spm: 4
Service: Mine drainage. Pumps in four lifts totalling 112 fathoms, raising 640 gallons/minute at 4 spm (in 1893)

Originally known as Sauchie Colliery, in 1792 it was taken over by the newly established Devon Iron Company, for whom it supplied iron ore and coal. The Neilson engine was erected in c.1854, but before the pitwork could be installed, production was stopped and around 1857 the Devon Iron Works also ceased operations. After an interval of 25 years, the pit was re-opened in 1879 by the Alloa Coal Company, and the engine, at the top of the downcast shaft, was fitted with pumps etc. In 1932 it was superseded by electric pumps, at some point most of the engine was scrapped (only the cast iron beam and the top section of the pump rod now remain), and in 1960 the colliery was closed. According to the report in the *Colliery Guardian* (Vol. LXVI, 1893, p.367), the engine was Cornish, 3-valved and condensing. (This may not necessarily mean that there were only two rather than three valves in the top nozzle, in addition to the exhaust valve below.) There were three plunger pumps – two 24 inch and one 14 inch – and one 14 inch bucket pump at the bottom of the shaft. The stroke in the shaft was 10ft (the beam being unequal). The beam, total length between centres 35ft 6ins (derived from measurement of the indoor section), with substantial catchwings to prevent overstroking, is an impressive relic, which will presumably continue to be preserved.

DUNBARTONSHIRE

205. Bonhill, Balloch Slipway (NS385824)

Photo taken/Reference: 20.5.1983/408-12
Type: Horizontal twin cylinder, driving slipway haulage winch
Maker and Date: John Bennie, Star Engine Works, Glasgow. 1902
Cylinder dimensions: c.12ins x 20ins. Slide valves; Stephenson reversing links
Hp: 35 **Psi:** 65 **Flywheel diameter:** 4ft 6ins
Service: Haulage of slipway carriage

Balloch was a terminus of the North British Railway at the south end of Loch Lomond, and the patent slip was built c.1900 by the North British Steam Packet Company, an offshoot of the railway company. Housed at the top end was the haulage winch, a cross-coupled slide valve engine, with three stages of reduction gearing (c.4.25:1; 5:1, and 5:1) to the 3ft diameter cable drum. Steam was provided by a large locomotive type boiler, pressure gauge red-lined at 65 psi – a replacement of the original vertical boiler. In recent years, it had been mainly used, seasonally, to pull the *Maid of the Loch* out of the water. But out of use from 1981, this vessel remained for a long period tied up at the pier and gradually deteriorating. In 1990, the slipway engine was used again with compressed air to pull up another vessel, the *Countess Fiona*. This resulted in the carriage, mainly wood but held together with lengths of rod and railway line, collapsing, and the damage was aggravated when the *Countess Fiona* was broken up on the slipway in 1999. However, on 27th June 2006, the slipway and its engine, with yet another replacement boiler (second-hand from a crane), having been restored, were used once again to haul the *Maid of the Loch*, now also being restored, out of the water.

203

204

206. Old Kilpatrick, Ministry of Defence, Old Kilpatrick Oil Fuel Depot – No.1 Pump House (NS468719)

Photo taken/Reference: 20.5.1983/407-10
Type: Horizontal duplex non-rotative pumping engine
Maker and Date: Worthington-Simpson Ltd., Newark. 1918
Cylinder dimensions: 22ins x 16ins. Slide valves
Spm: 30 (36 max) **Psi:** c.90 (150 max)
Service: Fuel oil pumping. Two double-acting plunger pumps (rams 11ins dia x 16ins stroke). Maximum capacity: 200 tons/hour, discharge pressure 250 psi

Of the four Ministry of Defence oil fuel pumping stations seen in the 1970s, this was the only one in which the main pumps (three here) were large duplex, but single-expansion engines. They had slide valves over the tops of the cylinders, operated in the usual way by the movement of the piston/pump rod on the opposite side; they exhausted to atmosphere at 25 psi. The pumps were used to boost the intake from tankers, for inter-tank transfers and for drawing from tanks for bunkering purposes. There were three oil-fired "Economic" type boilers, one at least (probably all) by Foster, Yates & Thom Ltd., Blackburn. A second pump house some distance away (the "Halls" Pump House) contained two inverted vertical single cylinder non-rotative pumps by J.P. Hall & Sons Ltd., Peterborough (cylinder 22$^{1}/_{2}$ins x 2ft stroke), each of about half the capacity of the Worthingtons. Here, there were another two Foster, Yates & Thom boilers, but there was also a steam line from the No.1 Pump House. By 1979, one of the Worthingtons was already disused, and all was closed down in October 1982, before this photograph was taken; probably all the engines remained until their scrapping in the early 1990s.

207. Old Kilpatrick, Scott & Sons (Bowling) Ltd., shipbuilders & repairers, Littlemill Slip Docks (NS442736)

Photo taken/Reference: 20.5.1983/408-34
Type: Inverted vertical twin cylinder, driving slipway haulage winch
Maker and Date: S. & H. Morton & Co., Leith. c.1890s?
Cylinder dimensions: c.12ins x 16ins. Slide valves; Stephenson reversing links
Psi: 100
Service: Haulage of vessels for repair up slipway. Maximum pull 1,000 tons

This interesting engine was another for slipway haulage, in this case with twin inverted vertical cylinders, cross-coupled and standing some distance apart, total height 7 feet 6 inches. All that is known for certain of its origin is the name of the maker, S. & H. Morton, cast onto the disc cranks. It may date from the 1890s, its previous location seems to be unknown, and it was said to have been second-hand here, installed during the First World War – according to one source in 1916. The drive to the haulage drum was via three stages of reduction gearing, with two alternatives for the first. (The approximate ratios were 2.5:1 or 5:1; 5.5:1; and 6:1.) There was a warping pulley on the third motion shaft. The boiler was a "Steambloc" gas/oil-fired boiler of 1958, with a normal steam pressure of 100 psi. Scotts used the slipway for the last time in December 1979 and closed the yard shortly afterwards. For a time in the 1980s, the premises were occupied by S.B. Offshore Bowling Base Ltd., and the slipway engine was retained, although it probably saw little if any use. This visit in 1983 seems to be the last record of the engine's existence – by 1990 S.B. Offshore had gone and the slipway engine had been scrapped.

FIFE

208. Aberdour, Cuttlehill Farm, Crossgates (NT156894)

Photo taken/Reference: 23.8.1979/181-10
Type: Vertical single cylinder
Maker and Date: Unknown
Cylinder dimensions: 8$^{1}/_{4}$ins x 18ins. Slide valve
Flywheel diameter: 6ft 5ins
Service: Driving threshing machine

This is a rare, if incomplete, example of a surviving farm threshing mill engine. The farm, previously part of the Wemyss Castle estate, in 1854 passed into the ownership of the Carron Company as an incidental, but apparently not completely unacceptable result of that company's acquisition of the mineral rights. (The Company's usual practice was not to acquire mineral rights alone, and as a result of this and similar purchases, they became farmers on a significant scale.) The engine, which could well be mid 19th century or slightly later, is located in a lean-to building against the barn which would have housed the threshing machine. It is of crude construction, the main bearing supported by the two iron columns which also form the steam supply and exhaust pipes. The second bearing is in the barn wall, beyond which the shaft, from which the drive was taken, has been cut off. There is a D-slide valve – the cover has been removed. Much is missing, including the connecting and eccentric rods, the piston rod guide and the governor, which presumably stood over the crankshaft and was driven by the bevel gear which remains. (Another anonymous example of a farm engine, also a vertical with two columns, but even less complete and now believed to have been scrapped, was seen at Mossburnford Farm, Jedburgh in 1975.)

205

206

209. Inverkeithing, Ministry of Defence, Rosyth Oil Fuel Depot (NT113820)

Photo taken/Reference: 17.5.1983/399-19

Type: Horizontal duplex tandem triple-expansion non-rotative pumping engines

Maker and Date: Worthington Pump Co. Ltd.*, London (built at Newark). Nos. 2997–8. 1917

Cylinder dimensions: 12ins/19ins/30ins x 2ft. Semi-rotary slide valves; HP cylinders each with two auxiliary cut-off valves

Hp: 366 bhp (trial–1917) **Spm:** 26 (average) **Psi:** 150

Service: Fuel oil pumping. Pump rams 12$^{1}/_{2}$ins x 2ft. Design capacity (at 26 spm): 250 tons/hour, discharge pressure 130 psi. (Capacity in use: 300–350 tons/hour.) *This was the holding company owning the shares of James Simpson & Co. Ltd.

In 1915–16 the Admiralty ordered a total of ten Worthington type triple-expansion engines for warship refuelling installations – eight of the same size: four for Rosyth (Nos. 2996-9) and four for Port Edgar (see under Torpoint – No. 37); and two slightly smaller for Corfe Mullen (Nos. 2979–80). One more for Lyness (No. 3105) followed in 1935. As at Torpoint, the Rosyth engines originally worked with condensers, but around the 1960s these were removed and the wooden cooling tower demolished. Also, apparently rather earlier than at Torpoint, the original Lancashire boilers were removed, superseded by two Foster, Yates & Thom twin furnace "Economic" boilers, dated 1953–5. Presumably it was then that the engines' working steam pressure (originally 125 psi) was raised. Two of the engines at Rosyth (Nos. 2996 and 2999) were the last such to continue in operation, finally stopping in September 1994. (The inner two, illustrated, had been stopped earlier and used for spares.) All were scrapped – presumably in the late 1990s. (Now, of the eleven Worthington triples referred to above, only three, preserved at Lyness, in the Orkney Islands, remain.)

208

209

210. Leslie, Smith, Anderson & Co. Ltd., paper manufacturers, Fettykil Mills (NO253015)

Photo taken/Reference: 26.8 1975/35-11
Type: Inverted vertical single cylinder
Maker and Date: James Milne & Son Ltd., Edinburgh. 1897
Cylinder dimensions: c.12ins x 15ins. Slide valve
Hp: c.35 **Rpm:** 100
Flywheel diameter: 5ft
Service: Drove DC generator for electric lighting

This engine was used for lighting, particularly in the area where the paper was finished and inspected, replacing the earlier gas lighting in the mill, and was presumably installed new. It provided far too little power and it was said that only a few years later, generators driven by two James Howden compounds were put in (these had evidently been gone some years). The date (1897) is based on that of the generator, Electric Construction Co. Ltd., Wolverhampton No. 4151, which was still preserved in the mill although off-site and not seen. A simple slide valve engine with Pickering governor, its location, at some distance from the boiler house, was inconvenient. The "steam hammer type" frame is very similar to that of the Tangyes engine at George Taylor's Bolton brass foundry (see No.86), but contrasting with this, the Fettykil engine frame provides support for only a single bearing, with a second in the wall alongside. Drive to the generator was by belt from the flywheel and the exhaust steam was said to have been used for space heating. As far as is known, the engine remains in situ today.

211. Leven, James Donaldson & Sons Ltd., Wemyss Sawmills (NO378006)

Photo taken/Reference: 26.8.1975/35-16
Type: Horizontal tandem compound
Maker and Date: James Carmichael & Co. Ltd., Dundee. 1923
Cylinder dimensions: 15ins/30ins x 3ft. Drop valves
Hp: c.400 **Rpm:** 75 **Psi:** 105
Flywheel diameter: 14ft
Service: Driving sawmill

This engine continued in regular use until November 1975, when it was the last working drop valve mill engine in Scotland. James Carmichael & Co., of the Ward Foundry, Dundee, was an important builder of a wide variety of engines, and now this is one of only two known to survive – the other is at Keathbank jute mill, Blairgowrie (see *Stationary Steam Engines of Great Britain*, Volume 2, No. 35). As will be seen, the governor controlled the inlet trip gear on the HP cylinder, which was behind the LP. The HP piston had a tailrod with a flat slide connected by rods to a double plate bell crank working the vertical air pump below. Drive was by ropes from the boarded flywheel (8 grooves, 7 ropes on). The single Lancashire boiler was fitted with mechanical hoppers, but these were said to be useless, and firing, with wood shavings etc., was by hand. With construction of a new mill, the life of the engine (by then stated by the company to be "almost worn out") was ended, and in 1981 it was removed to the National Museum of Scotland.

212/213. Tayport, James Donaldson & Sons Ltd., Tayport Sawmills (NO461286)

Photo taken/Reference: 18.5.1983/402-14 & 402-26
Type: Inverted vertical three cylinder compound (enclosed)
Maker and Date: W.H. Allen Son & Co. Ltd., Bedford. No. 23348. 1914
Cylinder dimensions: 15ins/17ins/17ins x 10$^{1}/_{2}$ins. Piston valves
Hp: 290 bhp **Rpm:** 350 **Psi:** 150
Flywheel diameter: 4ft 4ins
Service: Electricity generating plant for sawmill

This is a relatively rare example of a three cylinder compound (i.e. with one HP and two equal-sized LP cylinders), and one with an interesting history. With its generator, it was originally ordered by Sir W.G. Armstrong, Whitworth & Co. Ltd on 28th August 1912 for use as part of the electric generating plant in the Chilean battleship *Almirante Lattore*. However, the engine was completed to the requirements of the British Admiralty, for in 1914, soon after being commissioned, this vessel was appropriated by the British government for use by the Royal Navy (as HMS. *Canada* taking part in the Battle of Jutland). In 1920 she was returned to Chilean ownership, and again became the *Almirante Lattore*, continuing in use as the Chilean flagship until she was scrapped in 1959. When the generator set was bought by Donaldsons seems to be uncertain, but it was said to have supplied all of the Tayport sawmill's power requirements for nearly 40 years. Possibly it became surplus to the battleship's requirements at the time of a 1929–30 refit and was installed as part of a reconstruction of the Tayport mill which took place around 1932-3. The DC generator (200 kW, also by W.H. Allen) was direct-coupled and the engine exhausted to atmosphere or alternatively to process or heating. The engine remained on standby until 1977, after which the single Lancashire boiler was scrapped. The engine also was scrapped around 1986–7.

211

W.H. ALLEN, SON & Co Ld
BEDFORD, ENGLAND. 1914.
23348 No OF ENGINE
350 REVS PER MIN
15 & 17 & 17 DIAM OF CYLRS
150 STEAM PRESS
10 ½ STROKE
290 B.H.P.

212

210

213

LANARKSHIRE

214. Glasgow, Ibrox Baths (Glasgow District Council), Clynder Street (NS558651)

Photo taken/Reference: 28.8.1979/184-26
Type: Horizontal single cylinder
Maker and Date: Unknown
Cylinder dimensions: 3½ins x 6ins. Slide valve
Flywheel diameter: 2ft
Service: Drove agitator rakes in water filters

This tiny engine was part of the Baths and wash-house filtration plant, its purpose to drive the agitator rakes of the two patent pressure filters by Bell Brothers of Denton. (The fine sand in the filters had to be cleaned by periodic backwashing, during which the sand beds were agitated by the revolving rakes.) The engine was of the simplest possible design, with a looped frame supporting the two bearings and trunk guide, a bent shaft crank and overhung cylinder. The rakes were driven by gearing from a long overhead shaft, belt-driven from a pulley, only 8 inches in diameter, on the end of the engine shaft. Little information on the history of the Baths (or on the engine) has come to hand – doubtless it was a much-needed facility in this comparatively poor area. It has been suggested that possibly the engine as well as the filters were built by Bell Brothers. The engine had been out of use since around 1965, presumably succeeded by an electric motor drive. Still in use, however, was a 1958 Babcock & Wilcox water tube boiler which had replaced two Lancashires. By 1993 the future of the Baths was in doubt, and their closure seemed imminent.

214

215

215. Strathaven, Elder & Watson Ltd., knitwear manufacturers, Dunlop Street Works (NS700443)

Photo taken/Reference: 29.8.1975/37-12
Type: Horizontal single cylinder
Maker and Date: Reputedly Crawhall & Campbell, Glasgow. c.1865?
Cylinder dimensions: 9ins x 18ins. Slide valve
Rpm: 80 **Psi:** Up to 50 **Flywheel diameter:** 6ft
Service: Driving knitwear machinery

This pleasant little works, established by Elder & Watson in 1880, still retained much of its 19th-century atmosphere. No evidence as to the maker of the engine has come to light – dates of c.1850 and c.1865 have been suggested. It was said to have been installed here second-hand – could this have been around 1895, the date of the boiler? It has a flat bed mounted on a tall stone base (9 feet 7 inches in length), and all seems consistent with a mid 19th-century date, apart possibly from the detached trunk guide, which it has been suggested could have been a replacement. The Pickering governor also is unlikely to be original, and may have been put on when the engine was moved to Dunlop Street. The drive was by belt from a pulley (32 inches diameter) to a lineshaft, and among machinery driven from this and still present next door were a milling machine by James Melrose & Sons, Hawick and a hydro-extractor. The engine had been disused since 1927, but the small Cornish boiler, Wm. Wilson & Co., Lilybank Boiler Works, Glasgow No. 2406 of 1895, was still in use (showing a pressure of 20 psi). The works was closed in 1979 and later demolished; the engine is preserved at the Almond Valley Heritage Centre, Livingston.

MIDLOTHIAN

216. Edinburgh, Scottish & Newcastle Breweries Ltd., St. Ann's Brewery, Abbeyhill (NT269741)

Photo taken/Reference: 17.5.1983/397-35
Type: Diagonal single cylinder
Maker and Date: Murray & Paterson, Coatbridge
Cylinder dimensions: c.6ins x 10ins. Slide valve
Flywheel diameter: 2ft 6ins
Service: Driving hoist in maltings (maximum working load 3 cwts)

The engine and hoist, approximately 10 feet apart, with trap doors in the floor between them, were located at one end of the top floor of the maltings. The whole engine was mounted on a slanting framework on a flat base. The drive was by belt from the pulley (30 inches diameter) next to the wall to the hoist drive shaft pulley (18 inches diameter). From this, the drum shaft was driven via friction pulleys (8 inches and 30 inches diameter) and there was a winding pulley fixed to the ceiling above the trap doors. Nothing on the history of this engine and hoist has come to hand. It was presumably installed by Robert Younger Ltd. who founded the brewery in 1854 and carried on here until 1961, when brewing ceased after a takeover by Scottish & Newcastle Breweries Ltd. Possibly malting stopped at the same time, but in any case as the hoist had been fitted with electric motor drive, the steam engine may have been completely superseded before the end. Shortly after this photograph was taken, the engine was acquired by the National Museum of Scotland.

PERTHSHIRE

217. Auchterarder, Robert White & Co. (1929) Ltd., woollen & worsted manufacturers, Glenruthven Mills (NS955129)

Photo taken/Reference: 19.5.1983/403-37
Type: Horizontal tandem compound
Maker and Date: Maker unknown. 1873
Cylinder dimensions: $10^{1}/_{2}$ins?/20ins? x 30ins. Slide valves
Hp: c.70 ihp **Rpm:** 70 (1975) **Psi:** 40–60 **Flywheel diameter:** 10ft
Service: Driving weaving shed and other machinery – up to 120 looms (cut down to c.40 looms by 1975, 10 by 1980)

This is a most attractive if rather antiquated looking installation, hardly the most up-to-date power plant when installed c.1918. The mill was built for cotton weaving in 1877, and the engine, reputedly from a sawmill in the Glasgow area, was presumably put in as a replacement of a smaller engine. It was erected on a base of glazed bricks and surrounded by decorative wooden fencing. Both slide valves are operated by a single eccentric and there is a jet condenser alongside the HP cylinder, which is behind the LP. The air pump is directly underneath the condenser and driven from the engine crosshead by a side-rod which also extends forward to a boiler feed pump. The only speed control is by a Pickering governor. Drive was from a 6ft 6ins diameter pulley by a canvas belt to the weaving shed lineshaft. There was a small Cornish boiler by Penman & Co. Ltd., Caledonian Iron Works, Glasgow, hand-fired, built as late as 1949. The engine was stopped at the end of July 1980 and the mill closed completely a few years later – it is now used for other purposes. But the engine remains and indeed was steamed occasionally during the 1980s.

216

217

STIRLINGSHIRE

218. Grangemouth, BP Chemicals Ltd., Bo'ness Road - Methanol Plant

Photo taken/Reference: 16.5.1983/393-12

Type: Horizontal single cylinder rotative compressors

Maker and Date: Ingersoll-Rand Company, Painted Post, New York. Nos. BX2152 and BX2150. 1961

Cylinder dimensions: 21$^{1}/_{2}$ins x 30ins. Piston valve

Rpm: 140 **Psi:** 200

Flywheel diameter: 11ft 3ins

Service: Gas recycle compressors. Used for circulating synthesis gas around the methanol reactor

British Petroleum Chemicals Ltd., manufacturing chemicals from petroleum, had been formed in 1947. A number of American-built horizontal steam compressors were installed at Grangemouth by the company and its successor (from 1956) British Hydrocarbon Chemicals Ltd. These comprised four twin cylinder engines by the Worthington Pump and Machinery Corporation, Buffalo, 1951, in the Ethanol Plant; four more by Ingersoll-Rand Company, New York, 1956, in the No.2 Ethanol Plant; and the two single cylinder engines of 1961 seen here. All had ceased work – in 1982, 1981 and 1975 respectively and they were probably scrapped around 1984/5. The Methanol Plant engines were stated to be Ingersoll-Rand's Type XPV1/2. The compressors were 7$^{1}/_{2}$ins diameter and double-acting, with a rated capacity of 171 cubic ft/minute and a maximum working pressure of 5,200 psi. Fitted with tailrods, they stood at opposite ends to the steam cylinders and were presumably driven from the crossheads by side rods passing above and below the crankshaft – all was enclosed. The valve gear was similarly obscured by casing, but the piston valve, apparently with hand-adjustable cut-off, was doubtless driven by a single eccentric.

218

Makers Index

Managing at the Front Line

Second Edition

A handbook for managers in social care

Terry Scragg

Managing at the Front Line
A handbook for managers in social care
Terry Scragg

Published by:
OLM-Pavilion
Richmond House
Richmond Road
Brighton BN2 3RL
UK

Tel: 01273 623222

Fax: 01273 625526

Email: info@pavpub.com

Web: www.pavpub.com

A catalogue record for this book is available from the British Library.

ISBN: 978 1 84196 252 8

OLM-Pavilion is the leading training and development provider and publisher in the health, social care and allied fields, providing a range of innovative training solutions underpinned by sound research and professional values. We aim to put our customers first, through excellent customer service and good value.

Pavilion editor: Jo Sharrocks and Sanaz Nazemi, OLM-Pavilion
Cover and page design: Emma Garbutt, OLM-Pavilion
Cover image: Rodolfo Clix; www.sxc.hu
Printed on paper from a sustainable resource by Ashford Press, Southampton.

Updated and revised

Managing at the Front Line

Second Edition

A handbook for managers in social care

Terry Scragg

Managing at the Front Line: Contents

Managing at the Front Line: Acknowledgements

I would like to thank a number of people who have helped me at different stages in the preparation of this new edition, including, Gill Butler, Sue Craig, Charlotte Dawber, David Gaylard, Christine Knott, Simon Livesley, Rebecca Long, Beccie Mannall, Janet McCray, Andy Mantell, Bethan Marshall, Debbie Medlock and Tracy Vine. I would also like to thank the frontline managers and social care staff who participated in the group discussions and the respondents to the survey of management issues in social care services. Finally, my thanks to Joanna Sharrocks and Sanaz Nazemi at Pavilion Publishing for their editorial guidance and support throughout the process of preparing this book for publication.

Terry Scragg

Managing at the Front Line: About the author

Terry Scragg was formerly principal lecturer in management in the School of Social Studies, University of Chichester, where he was responsible for postgraduate courses for health and social care managers. He originally worked in social services departments as a mental health social worker, team manager and training manager before moving into higher education. He has undertaken a wide range of training and consultancy in social care services, both in the state and independent sectors, and has written numerous books and articles drawing on this experience. He has recently co-editored *Reflective Practice in Social Work* (2007) and *Safeguarding Adults in Social Work* (2008). Alongside academic and consultancy work he has had a long-term commitment to services for people with learning difficulties, including research, advisory and trustee roles. He has an MA in social service planning, a PhD in management and is a fellow of the Chartered Institute of Personnel and Development. He is currently a visiting fellow at the University of Chichester and a trustee of a learning difficulties service in West Sussex.

Managing at the Front Line: Foreword

Nearly every inspection by a regulatory body, report by a think tank, or inquiry investigation points to the need for better leadership and management at the front line. In the words of the author, '*Frontline managers play a critical role in social care services and significantly influence the quality of practice*'. Yet, as many managers state, their first management role is usually attained without much developmental experience and certainly without formal training.

Although no handbook can take the place of a commitment to development within organisations themselves, this is the ideal handbook for a practitioner who is stepping into a management role for the first time.

Following the successful publication of the first edition in 2001, Terry Scragg develops the approach across the board in a way that is admirably clear and accessible yet evidentially sound. I welcome the way this handbook considers the role of the manager, managerial techniques, developing the service and the development of the individual. The whole approach is grounded in practice with a clear vision for a better future.

Terry is a great teacher who lives his commitment to the values inherent in the social approach to health and social care. I always brought him into services that required radical and value-based change, and I was never disappointed.

This will add to the knowledge and confidence of junior managers, and thus lead to better outcomes for service users and carers. I commend it to you.

Peter Gilbert

Professor of Social Work and Spirituality, Staffordshire University
Chair, The National Development Team for Inclusion

Managing at the Front Line: Introduction

About this updated and revised handbook

This handbook is for managers who are working in a range of roles in social care services. What they all have in common is management responsibility for a team (or teams) of staff who are providing services directly to users and carers, either in the field, group care (day and residential services) or domiciliary care settings. The handbook has been written to help improve management skills, whether for a newly appointed manager, or for those who have been managing for some time, but feel they could benefit from reflecting on their management practice.

The first edition of this handbook was written in response to the needs of newly appointed managers in the statutory and independent sectors of social care; the accessible text was organised in a way that made it easy to refer to specific aspects of management. This second edition retains the overall aims of the earlier edition but has been updated and expanded to address a wider range of management issues, and has delved more deeply into some areas of management. There are, in addition, new areas that were not addressed in the first edition or where there have been new policy initiatives or legislation.

This new edition has also continued the approach used in the first edition of drawing directly on the experience of frontline managers and staff in influencing the content of the handbook, and providing examples of managing and working in social care services.

In revising the handbook, I would particularly like to mention the West Sussex Care Training Consortium who I have worked with for the last three years providing management training and development for staff in social care services. Working with many frontline managers has given me a valuable insight into the day-to-day demands on them, as they strive to provide high-quality services in a sector that continues to experience a range of pressures and constraints. Many of the examples used in the handbook come directly from this work. Without this dialogue with frontline managers and social care staff, who have attended Care Training Consortium courses, or who in addition responded to the survey of managers, it would have be difficult to write authoritatively about the experience of managers and their staff and the challenges that they face, in a way that I hope that readers will recognise.

How can I best use this handbook?

How you decide to use this handbook will very much depend on your needs. It is first and foremost a practical resource to support you and is not intended to be read from cover to cover in one go. It is something to dip into when you need to understand some aspect of management. It will help you improve your management skills or plan how you will manage a particular function in the future by reference to some of the principles and processes of management in a social care context. It focuses on a number of key areas of management that have been highlighted in consultations with a wide range of frontline managers in social care organisations and their staff, both in the statutory and independent sectors.

It should be stressed that a handbook of this nature cannot possibly cover all aspects of managing at the front line and some important areas of practice have been dealt with superficially. Should you want to explore a particular issue in more depth, then each chapter has references that can be followed up for further reading, or websites where you can access additional information on the topics covered in the handbook.

Section 1

On becoming a manager

Section 1

Chapter 1

Managing at the Front Line: Moving into management

Key points

- **It is common for managers to be appointed with little preparation for the demands of management and they can feel unprepared for the transition from the role of practitioner to frontline manager.**
- **Moving into management from a being a practitioner requires the development of new knowledge and skills. It can take time to learn about a new role and organisation, and the transition to management can be uncomfortable. It is normal to feel de-skilled for a period.**
- **It is necessary to let go of the old role before adopting the new one, acknowledging the value of practitioner skills, but recognising the importance of learning how to become effective as a manager.**
- **Time and a positive engagement with the management role and its demands can quickly lead to feelings of effectiveness as new knowledge and skills are gained.**
- **Identifying and asking for support (whether formally or informally) is important during the transition to the new role.**

Key words

Demands of new role; transition; new skills; adjustment; growth of effectiveness; support.

Introduction

Becoming a manager constitutes a significant change from a practitioner role as you learn to think and feel differently about your new role. Promotion means moving from a specialist care role where you perform specific tasks, such as social work or residential work, to one where you become more of a generalist involved in diverse tasks and, in turn, become identified with the management of the organisation and its strategies and policies. In the practitioner role you are mainly working on your own, for example, as a social worker, in co-operation with other professionals, or providing direct care in a group care setting. Once you move into the management role your main responsibility is to get things done through other people within the formal authority of management.

Most managers learn through experience what it means to be a manager as very few have any preparation or training for the management role. Research by Lawler and Hearn (1997) found that staff appointed to their first management post often had considerable doubts about their confidence to carry out the required tasks. This research is confirmed by the accounts of managers interviewed for this handbook who often felt unprepared for the management role and endured an uncomfortable period as they adjusted to the demands of a new job.

As Harris and Kelly (1992) observe, you may have been pushed or pulled towards the new job for a variety of reasons. You may have been dissatisfied with your previous job or attracted by the status and increased salary of a management post. Getting the job is often the high spot that can be quickly followed by a sense of anticlimax as the demands and the difficulties associated with the new job move into the foreground. These feelings are not unique to management roles as often the demands of a new job lead to a difficult transition period until confidence and competence in the new role are established.

New demands

With the move into management you are now required to develop an additional range of skills and knowledge specifically concerned with management. Although writers on social care stress the work of the practitioner as a solid grounding for management (for example, making decisions, influencing outcomes, seeking solutions to novel and complex problems) moving into management means making personal changes in a number of ways:

- you may have been promoted from within the same team, and now assume responsibility for the work of people who were previously your peers
- your relationship with your peers changes when you take on a management role as you assume managerial authority in relation to their work
- you have to adapt to expectations of your staff and your manager, so that you can act decisively in a wide range of situations, including meeting what can be the conflicting demands of management and practice
- you have to learn the language and norms of management and how to work effectively in a new organisation, or if you are in the same organisation, where there will now be different expectations placed on you and agendas that you are unfamiliar with
- you have to establish a working relationship with your manager and discuss what their priorities are in terms of your team's performance
- you become accountable for the performance of your team.

In the transition from a specialist practitioner role to the generalist management role, you will experience what Hill (2003) describes as moving from being someone who knows a lot about, and concentrated on, one thing (the practitioner focus), to being someone who knows something about, and works on, a lot of things (managing a team with its diverse agenda and issues). Success no longer depends solely on how well you do your own job, but on how well you get things done through other people. As a manager you will be judged on the performance of your team. See Box 1 for a summary of examples of the different roles.

Box 1 Summary of roles

Social care worker	**Frontline manager**
Works to perform a service	Ensures that the service reaches desired standards and is done in a timely fashion
Has a short time span for making decisions	Plans in terms of weeks and months
Follows the direction of others	Gives direction to others
Is accountable for own performance	Is accountable for team members' performance
Performs according to standards set by manager	Sets or interprets standards and measures the performance of others

The transition to the management role

The transition to the management role means that you can experience a major life change.

Although you may have felt confident about your practice before you moved to the new job, the challenge of the new role can lead to feelings that are common to many forms of transition, including personal crisis or enforced changes. Kakabadse *et al* (2004) have described how all managers go through phases in the process of entry and socialisation into a new role and organisation. Whether you are promoted within the same organisation or have accepted a job in a new organisation you will experience elements of this transition (See Figure 1 on p9).

During this transition period you have to learn what it means to perform effectively in the new role and this is a process that takes time. It is likely that you will experience problems and frustrations as you come to terms with the new role. How well you cope with the transition will depend on the degree of change demanded, your own capacity to adapt to the new role and the support and guidance you are given during the transition.

To some extent, all managers experience a similar transition, whether they are promoted within an organisation or accept a job in a new organisation. In essence, they have to learn to perform effectively in a new job and organisation and this process takes time (Kakabadse *et al*, 2004).

The words of this manager who moved to the third sector describes her reaction to managing in a new organisation:

> *'I was managing in a local authority before I took on this new job, but I still experienced difficulties when I moved jobs. It was getting used to a different culture from the local authority, different processes and different jargon. There were also new demands in learning to work with commissioners, which I hadn't experienced before.'*
>
> Team manager, voluntary organisation

Phase 1: Getting used to new demands and responsibilities

Whether you were promoted internally or changed organisations, your job content will be different. As you become familiar with your new responsibilities and new environment your effectiveness is likely to drop at first. The reactions you may experience at this stage can include feelings of being overwhelmed. Your level of knowledge and understanding may mean you are unable to meet expectations, and the way that things are done in the organisation may be unfamiliar. You may feel that you can make very little contribution at this stage. Once you become more familiar with the new environment – ideally with the support of your own manager and your team colleagues and a planned induction programme – you will begin to feel your old sense of self emerging.

Phase 2: Relearning

When you have begun to become familiar with the new environment and some of the work demands, you have to start learning about the new responsibilities and job skills that accompany the role of manager. You will also have to develop or adapt your relationship with colleagues and other managers, and begin to understand their way of doing things and the expected norms of management behaviour in the organisation you have joined. This requires a rapid learning curve and can be accompanied by feelings of loss as you realise that you can no longer use many of your old skills.

The skills and knowledge developed in the practitioner role provide an important resource, but you will be faced with situations where you need to learn a new range of skills and this can affect your confidence. This stage of the transition can lead to feelings of panic and helplessness, although these feelings are the first step in relearning as you realise you need to let go of the past and develop the skills that will enable you to work effectively in the management role.

Phase 3: Becoming effective

Becoming effective in the management role means you have successfully negotiated the transition. You have let go of your old role and accepted the new reality and challenges and become more proactive. You can now start to experiment with new behaviours, new ways of working and new approaches. Slowly you will begin to incorporate the important aspects of your old role alongside the new ways of working.

You have now consolidated your position in the frontline management role and feel a growing confidence to deal with the demands made on you. You are beginning to see yourself more positively and there is a growing satisfaction with the new role.

Figure 1 Transition to a management role diagram

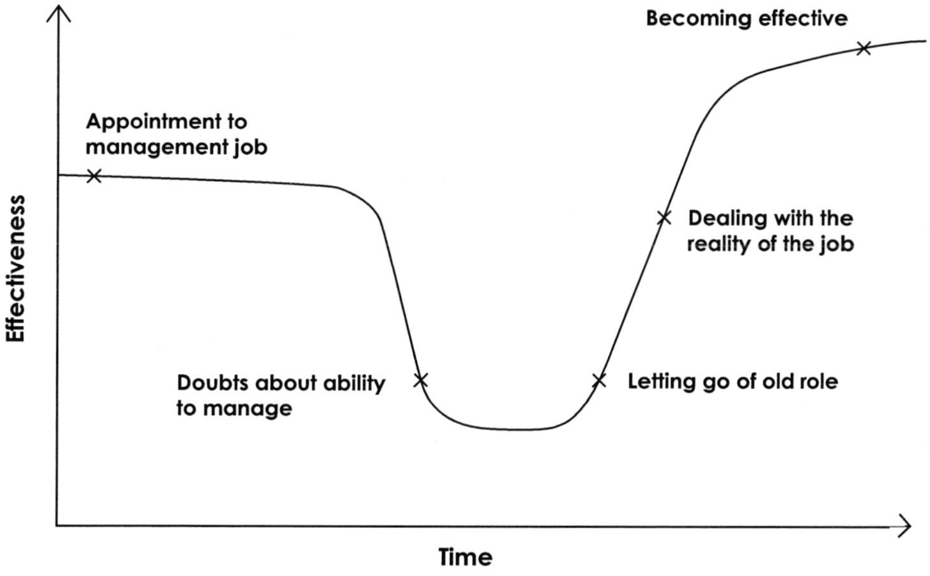

Adapted from Harris and Kelly (1992) and Kakabadse *et al* (2004).

Different reactions to the transition

It is important to stress that not everyone who moves into a management role will have the same reactions as those described in the process of transition. How quickly you adjust to the new post will be dependent on a range of factors, including: your feelings about the change of role, an awareness of the demands of the management role, the support and guidance provided by the organisation through induction and support from a line manager, and the rapport established with team members.

The transition to the management role takes time, with Hill (2003) describing managers taking until the middle of their first year in the new role to adapt to the demands of management. The new managers in Hill's study were then becoming aware of the pattern of their behaviour, had adopted a management style or at least a preferred way of working. In turn the new managers were slowly acquiring the management character that enabled them to effectively manage their team, including self-confidence, willingness to accept responsibility, patience, empathy and living with imperfect solutions. Of all these characteristics, self-confidence was seen as the most important, as this increased self-assurance and meant that managers were more open and less defensive in their dealings with colleagues.

> *'After I was promoted I felt de-skilled initially. My confidence dropped and I didn't feel I could contribute as a manager.'*

Team manager, social services department

Changing relationships with staff members

As you become established in the management role you are now in a hierarchical relationship with members of your staff team and you need to acknowledge this. Accepting this change is an important part of the transition and one that managers often describe as difficult to adjust to, particularly as they see themselves as team colleagues, but also as manager. This is often further complicated by promotion within the same team. A number of managers interviewed confirmed the view that it is necessary to accept a managerial identity and the responsibilities that go with that identity. This is a key change in the transition from practitioner to manager and can prove to be an identity crisis for some managers.

> *'The crunch came one day soon after I became a team manager. I knew I had to rethink my role and my relationship with the team. It was part of becoming a manager. I couldn't be everybody's buddy anymore. I'm responsible for their work now and I've had to think myself into the new job and acknowledge the responsibility that went with it.'*
>
> Team manager, social services department

Who can help you manage the transition?

In the past it was common for managers to be 'thrown in at the deep end' without proper induction, briefing and support, as reported by several managers interviewed.

To avoid this situation it is important to identify both formal and informal supports within the organisation. Other managers who will have experienced a similar transition will be able to empathise with you and suggest ways of coping. Your own line manager has a major responsibility here to support you and identify development opportunities that can improve your skills (although some managers I interviewed did not feel confident to be open about their support needs with their line managers and others described an organisational culture where asking for support was seen as a weakness).

> *'It's important to establish what your responsibilities are and to have clarity about your boundaries and lines of responsibility, and the expectations of your manager and your staff. You can't begin to work effectively until you are clear what your "fit" is with your boss and the team.'*
>
> Team manager, social services department

When you are developing your management skills it is important to share your feelings in an open and frank way so that you enable others to support you. Discussion with other managers can help you make sense of the process of transition. There are risks for managers in that they can become isolated when they feel under pressure to appear competent both with staff and other management colleagues. This is not healthy for the individual or the organisation and is one of the less attractive consequences of the 'macho management' culture found in some social care organisations.

Organisations that understand the relationship between individual and organisation effectiveness provide a more structured approach, including induction and enhanced support, initially from a line manager. Other approaches include the appointment of a mentor who periodically meets with the newly appointed manager to review progress and is available for consultation during the period of transition.

Conclusions

The transition from a practitioner to management role can make considerable demands on the frontline manager as new skills have to be learnt quickly during the early months in the role. The demands of the transition can mean that frontline managers feel uncomfortable as they adjust to the new role and slowly increase their effectiveness. The availability of support is important in helping the newly appointed manager adjust to the demands of the role and organisation.

References

Harris J & Kelly D (1992) *Management Skills in Social Care: A handbook for social care managers*. Aldershot: Ashgate.

Hill LA (2003) *Becoming a Manager: How new managers master the challenge of leadership*. Boston, MA: Harvard Business School Press.

Kakabadse A, Bank J & Vinnicombe S (2004) *Working in Organisations* (second edition). Aldershot: Gower.

Lawler J & Hearn J (1997) The managers of social work: the experiences and identification of third tier social services managers and the implications for future practice. *British Journal of Social Work* **27** (2) 191–218.

Managing at the Front Line: Changing roles and responsibilities

Key points

- **Frontline managers play a critical role in social care services and significantly influence the quality of practice.**
- **The scope of their management activities has increased as more responsibilities have been devolved to the front line.**
- **Their responsibilities encompass strategic, operational and professional levels.**
- **Supporting staff is central to the frontline role.**
- **A series of key activities that frontline managers need to engage with are suggested that can lead to greater effectiveness in the role.**

Key words

Changing management role; diverse demands; arbiters of practice; supporting staff.

Introduction

Since the first edition of this handbook there has been a growing recognition of the vital role played by frontline managers in social care. The importance of their support for standards, supervision of staff and ensuring that service users can fully engage with services is increasingly recognised and the key to effective delivery of services. It is gratifying to witness an increasingly wide range of statements that highlight the pivotal role played by frontline managers in social care services. A statement by the former Chief Inspector of Social Services sums up the growing awareness of the importance of the frontline manager's role.

> *'Frontline managers are the keystones of any social services organisation. Their quality and competence makes a significant difference to its performance. They manage the primary tasks and activities of the organisation. They have a key role in determining whether standards of practice are consistently maintained. They support staff engaged in complex, personally demanding practice, and ensure staff are continually developed in knowledge-based practice.'*
>
> Platt, 1999

What is frontline management?

Although the role of the frontline manager may vary according to the requirements of the employing organisation and particular responsibilities of a post, there are common features in all frontline management posts. All frontline managers have the authority and responsibility for planning and controlling the work of a group of staff through close contact. They are responsible for implementing policies at the point where it impacts on people using the service, and typically include team managers in community-based services and managers of group care (residential and day services) and domiciliary services. In this sense they are the visible managers of the organisation, and often the public face of management in an organisation.

Due to their influence in supporting and supervising staff and ensuring the development of high-quality practice, the frontline manager plays a crucial role in the success of the organisation and consequently the quality of services provided. Frontline managers can make or break top management plans due to their crucial role in the implementation of organisational strategies, which gives some sense of the importance of the role as stated by the former Chief Inspector of Social Services.

What is expected of frontline managers?

The role of frontline managers has changed significantly, along with wider changes in public services and the rapidly growing independent sector. These changes have resulted in more devolved management structures and greater responsibility falling on frontline managers for the implementation of a range of government initiatives concerned with improving standards of performance and translating policies into practice.

In research undertaken with frontline managers in a range of different agencies, which examined their tasks and roles, Henderson and Seden (2003) found that employers want them to manage the dilemmas, constraints and challenges that face services with three broad areas of work – operational, strategic and professional.

- **Strategic:** included using their knowledge of their local area in support of strategic planning, responding to new policies and initiatives and developing services alongside partners in other agencies.
- **Operational:** included focusing on the day-to-day support and supervision, work allocation, developmental and enabling role with team members. In a single day a manager could be dealing with a myriad of tasks in relation to staff members that underpin the delivery of services.
- **Professional:** included managing staff who deliver care services, including supporting staff who were dealing with complex issues where the frontline manager's knowledge of legislation and practice was seen as invaluable, and increasingly, managing multidisciplinary teams and helping staff focus on a single system of care.

This research exposed the tensions in the frontline management role, particularly between the operational and strategic demands on their time, and structural attempts to address these pressures through the introduction of senior practitioner posts to oversee day-to-day operational matters, although these were the areas that attracted managers to the role in the first place. This research confirmed the view that frontline managers play a key role in ensuring service users, staff, departmental and government needs are met. This is an overwhelming expectation, with a constant increase in demands on the role, in the context of continual government initiatives and the need to manage change.

Frontline managers' tasks identified by practitioners

A helpful approach to exploring frontline managers' tasks was explored in the National Institute for Social Work's (NISW) Management of Expertise project, which examined the conditions that support good social care practice. The project asked 'what approaches best enable staff to develop and sustain their practice expertise?' Its findings concluded that frontline managers hold a pivotal role as arbiters of practice standards and are the 'keystone' of the organisation, and that an effective frontline manager should be able to:

- manage practice, including supervision of individuals and the team
- manage workloads, workflow, service development and the administrative and financial systems to support these
- offer supervision that takes into account the emotional impact of the work and its effect on staff and the organisation
- understand the standards required for practice and integrate these into the work of the team
- contribute to the professional development of team members, both collectively and individually
- model and describe a professional approach to work, include appropriate use of authority, autonomy, responsibility and accountability, and expect and enable staff to understand and demonstrate this in their own work.
  (Kearney, 2004: 105–106)

What social care staff want from their managers

Discussion with social care staff from a range of services suggested the following expectations of their frontline manager:

- *'encouragement, guidance and support over my workload'*
- *'advice about specific cases'*
- *'regular supervision'*
- *'frequent team meetings to discuss issues'*
- *'an open door policy'*

- *'support the decisions I make'*
- *'feedback on my performance delivered in a constructive way'*
- *'support my development'*
- *'approachability and trust'.*

In making these statements staff recognised that it was difficult for their managers to always respond to their needs for a range of reasons, which confirm the findings of the research by Henderson and Seden (2003) of the multiple demands on frontline managers. Social care staff added:

- *'I am aware of the increasing responsibilities of my team leader who is less present in the team, which is frustrating at times when you need urgent support or decision-making.'*
- *'I would prefer that my supervision sessions were not interrupted by phone calls from senior managers.'*
- *'I don't think my manager is as supported as she could be by her manager.'*
- *'My manager seems to attend lots of meetings and is absent on many occasions.'*

Twelve steps to effective frontline management

Reflecting on the findings of research into the frontline manager role and their activities in Rosen (2000), Statham (2004) and Henderson and Seden (2003) and the response of managers who took part in discussions for this handbook, there emerges a number of key activities that constitute the main focus of the frontline manager's role.

1. Clarity about the goals of the service

You need be clear about the purpose of your service. It is from the organisation's statement of purpose (or mission statement) that the strategies and policies are developed. The activities of managers and practitioners are designed to contribute to the achievement of organisational strategies that stem from the statement of purpose. An understanding of the purpose of the service can also help you to communicate the 'bigger picture' to team members so that they understand what the organisation is trying to achieve and how their contribution fits in with this bigger picture. Securing agreement on how individuals and teams can work to achieve organisational goals is an essential part of the frontline manager's responsibility. Without this agreement it is unlikely that staff will feel their contribution is valued, and less likely that they will feel motivated to work to achieve organisational goals.

2. Knowledge of the social care task

Frontline managers are usually recruited from the ranks of practitioners, and as such bring to the role a detailed understanding of practice. They also bring the skills to deal promptly at first hand with problems arising at the point of service delivery, in order to direct, guide and advise practitioners as appropriate. It is the detailed understanding of the work of practitioners that is important in enabling frontline managers to make appropriate decisions and respond to staff as they struggle with the daily demands and contradictions of practice.

3. Establishing and maintaining quality standards

Frontline managers play an important role in designing and developing quality systems that establish and maintain standards of practice in their service (Hafford-Letchfield, 2007). They have both a quality assurance function, where service delivery is designed to achieve high quality, and quality control, which involves monitoring activities and comparing them with standards set and taking corrective action where necessary.

4. Managing performance

Closely related to the establishment of quality standards is the increased emphasis on managing performance, which is primarily concerned with aligning the strategies and objectives of the organisation with the work of practitioners and other staff. The growing emphasis on performance management means that frontline managers need to lead and motivate staff to achieve the high standards of practice that are a requirement of all services. This requires an understanding of both the qualitative and quantitative indicators for the service, and to be able to measure and evaluate the performance of the service being provided against these standards. This also means a willingness to taking action when performance falls below acceptable levels and the work of individual staff is unsatisfactory.

5. Managing change

The range and pace of change in social care organisations has increased significantly in recent years, with a plethora of government policy initiatives that have resulted in frontline managers having to take responsibility for the detailed implementation of these policy initiatives. Similarly, quality systems and performance management inevitably demand change, and can be particularly challenging for managers who need to develop skills in the sensitive management of change, where staff may already feel overwhelmed by demands on their time.

6. Being a link between the top of the organisation and the front line

Frontline managers are located at a point in organisations that is critical for the effectiveness of the service. They are the channel for the transmission of information from senior management to the front line, and in turn they transmit information up through the organisation so that middle and senior managers have a clear sense of what is happening at the front line. In playing this role, the frontline manager is the 'bridge' between senior management and the needs of frontline staff, interpreting the goals of the organisation to practitioners. Communicating and explaining management decisions is critical to the success of an organisation, and this in turn provides information where policies are not working and can represent the opinions of frontline staff to senior managers.

7. Holding together the different worlds of social care

Frontline managers are in a key position to hold together what can seem like different worlds, integrating the work of practitioners with the wider activities of the organisation and avoiding the split that can develop when frontline managers fail to achieve the integration between the work of practitioners and the demands of the wider department and, ultimately, government (Rosen, 2000). The increasing demands on frontline managers mean that their traditional orientation towards practice has been partly undermined by the dominant focus on the management of resources. If this tension is not understood and discussed in the team, it can lead to a split in identity and loyalties. It is the bridge between the front line and management that avoids the danger of staff identifying completely with either practice or management and neglecting the need to relate wider organisational needs to the realities of practice.

8. Managing and working with other professionals and services

With increased inter-agency partnerships pursued as part of government policy for all public services, frontline managers are increasingly managing and working in multidisciplinary and multi-agency settings, collaborating with a range of partners now part of the daily experience of professional practice (McCray, 2007). This means that frontline managers need to develop skills in both managing and working with a range of professionals that require an understanding of the different value bases of the professions and their unique contributions.

9. Staff as the key resource

Social care staff are an organisation's main asset, and the deployment of this important resource is the main responsibility of frontline managers. Management is primarily about getting things done through people consistent with the organisation's strategies and policies. This means that managers are responsible for co-ordinating work activities based on service priorities, guiding and supporting staff and monitoring performance.

To achieve these objectives, management is primarily a social activity where sensitivity and understanding of members of staff is essential in order to match their knowledge and skills with the service demands and requirements.

10. Being available to staff

Frontline managers operate in the 'here and now' for much of their time. This means being available for immediate advice and consultation on a day-to-day basis, identifying problems, responding rapidly to the concerns of staff and initiating or supporting corrective action as required. The complexity of the social care work means that frequent recourse to the manager will always be needed by some staff. The maxim – 'management by walking about' – is important here as it describes a key aspect of effective leadership which is visibility and keeping in touch with staff.

11. Supporting staff through supervision

As services become ever more responsive to the needs of service users and also focus their scarce resources on the most needy, it is increasingly recognised that supervision plays a crucial role in developing and monitoring practice standards and supporting staff, where the norm can be daily contact with distress and loss (Kearney, 2004).

Similarly the knowledge of practice is also important when providing supervision.

The practitioner's reflection of their work with service users benefits from the manager's understanding of the complexity of practice situations. All this can make considerable demands on the frontline manager who has to manage practice in all its diverse forms and requires a breadth of skills to supervise staff effectively (Morrison, 2001).

12. Developing a commitment to work-based learning

Finally, the frontline manager has an important role in facilitating and providing work-based learning for their staff (and not forgetting your own development). Formal education and training play an important role in developing professional expertise, but most learning is rooted in the workplace, arising naturally out of the demands and challenges of the work itself (for example, solving problems, improving quality or coping with change through interactions in the workplace with colleagues and users of services). The quality of management is highly significant, particularly the potential for creating a climate that facilitates work-based learning and contributes to the development of the service (Eraut, 1999).

Conclusions

The importance of the frontline management role is now firmly recognised. The multiple demands on frontline managers means that they have to understand the requirements of a diverse range of activities including strategic, operational and professional areas of work. The role is seen as critical as frontline managers are located in the point in services where they can directly influence practice. Their role in supporting staff is central to their role.

References

Eraut M (1999) Managers hold the key to developing knowledge and skills. *Professional Manager* **7** (2, March) 41.

Kearney P (2004) First line managers. The mediators of standards and the quality of practice. In: D Statham (Ed) *Managing Front Line Practice in Social Care*. London: Jessica Kingsley Publishers.

Hafford-Letchfield T (2007) *Practising Quality Assurance in Social Care*. Exeter: Learning Matters.

Henderson J & Seden J (2003) What do we want from social care managers? Aspiration and realities. In: J Reynolds, J Henderson, J Seden, J Charlesworth & A Bullman (Eds) *The Managing Care Reader*. London: Routledge.

McCray J (2007) Reflective practice for collaborative working. In: Knott C & Scragg T (Eds) *Reflective Practice in Social Work*. Exeter: Learning Matters.

Morrison T (2001) *Staff Supervision in Social Care* (Second edition). Brighton: Pavilion Publishing.

Platt D (1999) Preface. In: P Kearney. *Managing Practice: Report on the management of practice expertise project*. London: National Institute for Social Work.

Rosen G (2000) (Ed) *Integrity, the Organisation and the First-line Manager*. London: National Institute for Social Work.

Statham D (2004) (Ed) *Managing Front Line Practice in Social Care*. London: Jessica Kingsley Publishers.

Managing at the Front Line: Becoming effective in the management role

Key points

- **Becoming effective is first about understanding what is expected of the frontline manager role.**
- **Frontline managers need to undertake a series of priority activities during the first few months in the job, which can directly influence their longer-term effectiveness.**
- **There are demands and constraints on frontline managers but they have choices which, if identified and developed, can increase effectiveness. These choices should be related to service priorities.**
- **Positive views about the demands of the management role can build self-confidence and can help the frontline manager cope with the demands of the role and support effectiveness.**

Key words

Priorities; influencing factors; choices; self-confidence.

Moving into a management role in social care services increasingly means co-ordinating a range of diverse activities in the context of wider policy changes, with multiple demands and expectations in terms of the goals you must achieve. Becoming more effective in your frontline role is measured by the extent to which you achieve the targets you set yourself and those identified by your manager. This can only be achieved if you are clear about what is expected of you and therefore set yourself priorities that meet the needs of the organisation and members of your team, and focus on those activities that meet corporate aims.

What you can do to help you increase your effectiveness

In her book on managing for the first time, Mill (2000) describes the newly appointed manager's first 100 days as the honeymoon period. This is when managers need to have accomplished a number of important tasks that will help them develop their effectiveness. I have drawn on her ideas as the basis for the following six steps in setting out your priorities.

1. Know your organisation

If you are new to your organisation, or have been promoted internally, it is important to know what is expected of you in your role. This stems from an understanding of the corporate strategies for the service and its strategic objectives and how these relate to your team objectives. This will enable you to interpret them, communicate them to team members and implement them more effectively with the support of the team. Understanding strategy will also help you in deciding which tasks are a priority, what you should be achieving in wider organisational terms, and what you will be judged on in performance terms as a frontline manager.

2. Get to know your manager

You need to meet with your manager as soon as possible as he/she is one of your most important relationships and is best placed to help you map out what you need to achieve in terms of priorities. Your manager will have expectations of you in terms of performance and what you need to discuss with them before making decisions. Frequent meetings with your manager are important as these give you the opportunity to gain a wider understanding of the needs of the organisation and to provide information about the performance of your team. It is also an opportunity for supervision and consultation, mutual problem-solving and feedback on your own performance.

3. Communication across teams and organisations

It is also important to understand that you have an influence when you communicate with others, particularly across the range of professionals and teams that you will find yourself working within your management role. You need to create trust through the way you communicate with others as this will demonstrate effective leadership, particularly when there are difficult decisions to be made or at time of major change. Trust is achieved through open, comprehensive and candid communication (Wheeler & Grice, 2000) as it is easy for trust to be lost where there is inconsistency and dishonesty in communication.

4. Meet with your team

An effective manager is someone who knows their staff well. Ensure that you take the first opportunity to meet them. They are likely to be concerned about how you will manage the team and whether you intend to change things. An initial meeting where you can introduce yourself and learn something about the team members can be a good start. Use this meeting to discuss the main issues facing the team and what they feel the priorities are and what needs immediate attention. You can also explain your immediate priorities and importantly what you will not be doing until you have had an opportunity to discuss issues in greater depth with the team and with your manager. Be careful not to promise things you cannot deliver or for which you need the agreement of your manager as this can dent your credibility before you get a chance to demonstrate your effectiveness.

5. Meeting staff individually

This is an opportunity to learn about the individuals who make up your team and what is currently exercising them. You need to know about their individual responsibilities and what the key challenges are in their role. It is also important to explore whether they feel things could be improved and why. You will begin to get to know them as people and what their interests are, what motivates them, and about their career interests, including current training needs. Through this meeting and others that you arrange, you will be able to manage them in a way that they find supportive.

6. Developing your relationships with key colleagues and service users

As a newly appointed frontline manager, you will be aware that all staff working in social care services are in a customer relationship with colleagues focused around the delivery of a service. It is the effectiveness of this interdependent relationship that is an important measure of the quality of a service. In order that you are able to improve the responsiveness of your part of the service, you need to understand the needs, wants and expectations of colleagues, service users and customers. An important first step is to meet with colleagues who both provide services to your team and to whom you provide services. Use the opportunity of meeting to explore what works well, what would improve the service from their perspective, and what they would like to see in the future.

Other factors that influence effectiveness

In addition to the above factors, Martin and Henderson (2001) caution that effectiveness in the management role is not just a matter of having the right skills, as the reality of organisational life is more complex than that. They list five factors that can influence your effectiveness, not all of which are under your control.

1. The first factor is you; including the blend of skills, knowledge, attitudes, values and experience you bring to your job that all influence your effectiveness to some extent.

2. The second factor is that your job, which although it will have much in common with other management jobs, is likely to be unique, both in the detailed features and the unique demands that it makes on you. It is the match between your skills and the demands made on you which influences your potential effectiveness.

3. The third factor is the resources you have to work with, which as a frontline manager in social care will mainly be the members of your staff team. It is here that statements such as 'managers get their work done through other people' place the emphasis firmly on the importance of your team in enabling you to achieve your priorities. It is in the management of staff that you are likely to face your biggest challenge in terms of creating conditions where they are motivated and experience job satisfaction, which in turn leads to effective outcomes in terms of the overall service.

4. The fourth factor is the organisation. Here the structure and culture will influence what you are able to achieve as a manager. Your position in the structure of the organisation, and the nature of your authority and your responsibilities will dictate what you are able to achieve. The organisational culture, with its norms and standards, visible and invisible elements and ways of working will all influence your effectiveness.

5. The final factor is the external world outside your organisation, including the social, economic and technological changes that impact on social care, but also those organisations that your service interacts and collaborates with, and in formal partnerships, in order to deliver improved services.

Although there are a wide range of factors that influence effectiveness, it is important to understand that you are also an active element that influences what can be achieved. You are not a passive bystander whose activities are completely shaped by others, particularly if you have a clear view of what needs to be done and how you intend to achieve it.

Establishing priorities: your perception of what needs to be done

Once you have met with your manager, your team and other key colleagues and service users, you should have a clearer picture of your priorities. Similarly, you should now have a wider appreciation of the factors that influence your effectiveness, which will give you a much more realistic understanding of the complexity of the manager's world and what you need to take into consideration when establishing those priorities.

For example, making immediate but limited improvements to a service but also recognising that long-term improvements are necessary. Where priorities are concerned with immediate operational activity this needs little lead time, whereas budgetary priorities require a longer lead time and, finally, decisions concerning changes to staff attitudes, abilities and working relationships take the longest lead time of all.

Taking control of your activities

In order to develop your effectiveness in the frontline management role you need to take control of your activities, particularly your time, in order to work towards the achievement of the objectives for your service. This requires you to think about which activities will contribute most to the achievement of the objectives, and managing the time pressures that result in you dealing with situations in a superficial manner.

Demands, constraints and choices – the differences in what managers do

In research that examined what managers specifically do in their job, Stewart (1997) found that they emphasised different aspects of their work. Stewart's research revealed that managers in similar jobs differ in what they do, with each manager doing the job in their own way. She believes that this is because managers see their job differently and individually managers focus on particular aspects and ignore others. Managers also have different skills, and this leads them to give more time to the things they are good at, or interested in, and this influences how they spend their time. Stewart has described this as the demands, constraints and choices that all managers have in relation to their role.

- Demands: these are the 'must dos'– the levels of accountability that must be achieved – and usually refer to meeting your organisation's expectations. In particular, they refer to meeting the expectations of your line manager and staff about what should be done. Further demands can arise from the expectations of people outside the organisation, for example staff in other services with whom you are working in partnership.
- Constraints: these are the factors, both internally and externally, that limit what you can do. These can include: resources, legal requirements, policies and procedures and others' attitudes and expectations.
- Choices: these are the opportunities that you have to decide what you want to do and how you do it. This is the area where you can emphasise some aspects of your job and give lower priority to, or ignore, other aspects. As a first line manager you are likely to find your scope for choice is somewhat restricted, but you will still have scope for making choices about what aspects of your role you will develop.

Figure 2 Choices diagram

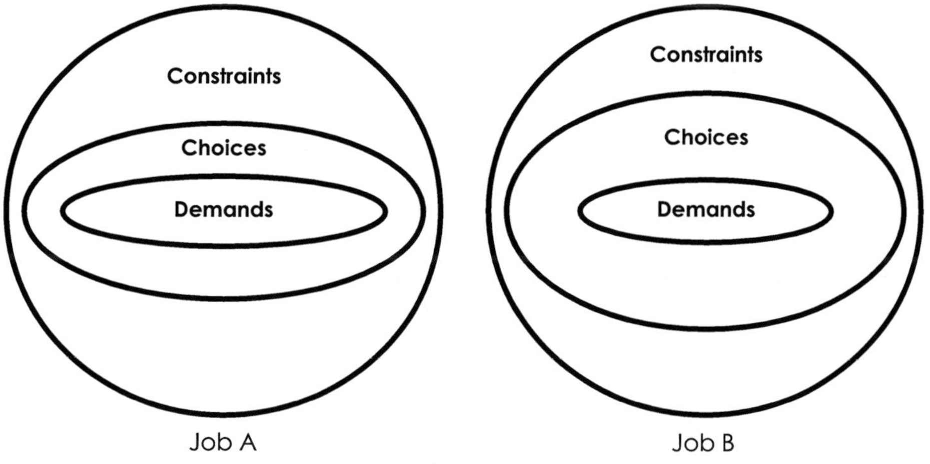

Adapted from Stewart (1997).

Job A illustrates a situation where the manager has to meet the demands with few choices within the constraints imposed. Job B illustrates a situation where the manager is successfully using their discretion to make the most appropriate choices about what to do and how to do it within the constraints.

It is important to remember that demands may be only perceived constraints. Using this framework of demands, constraints and choices you can analyse and understand your job, review how you approach it in order to take an objective and strategic view of it and assess your effectiveness.

All frontline management jobs are different and within each there are choices in both the content and design that a manager may fail to understand, and consequently end up performing inefficiently and ineffectively and feeling dissatisfied (Kakabadse *et al*, 1988).

You have more choices than you think

Commenting on Stewart's research, Cunningham (1994) suggests that the less successful manager is likely to see themselves as having less choice than the outside observer might identify, as seen in the diagram where one circle has much larger areas of choice than the other. This can be where managers make assumptions about the demands and constraints in their situation and fail to test them. The more successful manager tends to see more choices, which Cunningham sees as linked to feelings of autonomy and willingness to push the boundaries of the job. The successful manager is likely to have a good 'map' of the 'territory' in which they operate and work in a culture that encourages them to break out of any self-imposed straightjacket.

A similar argument is put forward by Hearn *et al* (1992) in terms of developing the culture of a service. They argue that a manager's behaviour affects the culture of a service and relate this to Stewart's assertion that the key difference in approach is between managers who see the job as primarily *responding to and coping with* problems and demands as they arise, and those who see it as operating in an environment inside and outside the organisation in which the manager tries to *control and develop and to mould* the culture they wish to create. It is possible for managers to exercise choice and control over their activities and this is crucial to the implementation of change or the development of innovative practice.

The 'inner game' of management

Finally, in this chapter on effectiveness in the management role it is useful to reflect on some ideas from the influential management thinker Drucker (1974). Drucker described one of the most important aspects of effectiveness in establishing yourself as a manager as being to do with what he describes as, the *inner game* of management.

The *inner game* is about managing your own self-esteem so that you derive satisfaction from undertaking management activities, including feeling comfortable with the authority of the management role and the responsibilities concerned with controlling staff and making difficult decisions. You also need to be able to manage the need to be liked in a way that does not interfere with performing the management role effectively, particularly where you have to make difficult decisions about resources and disappoint staff when you have to refuse their requests.

This contrasts with the *outer game* which is much more straightforward, but no less demanding. This encompasses all the tasks and activities of management, alongside the skills that characterise effective management such as allocating work, ensuring high standards of practice and supporting the work of individuals and teams.

Conclusions

Becoming effective in the frontline management role is essentially about understanding what is expected of you. There are a number of key activities that can improve your effectiveness in the first months in the job. You also need to be aware of the wide range of factors that can influence your effectiveness. You have many demands placed on you and you are constrained by a range of factors, but you have the opportunity to make choices that can increase your effectiveness and also give you greater job satisfaction. To be effective is also importantly about managing yourself, your emotions, your self-esteem and how you see yourself in the management role.

References

Drucker P (1974) *Management: Tasks, responsibilities, practice*. London: Heinemann.

Cunningham I (1994) *The Wisdom of Strategic Learning: The self-managed solution.* London: McGraw-Hill.

Hearn B, Darvill G & Morris B (Eds) (1992) *On Becoming a Manager in Social Work.* Harlow: Longman.

Kakabadse A, Ludlow R & Vinnicombe S (1988) *Working in Organisations*. Harmondsworth: Penguin Books.

Martin V & Henderson E (2001) *Managing in Health and Social Care*. London: Routledge.

Mill C (2000) *Managing for the First Time*. London: Chartered Institute of Personnel and Development.

Stewart R (1997) *The Reality of Management* (third edition). Oxford: Butterworth-Heinemann.

Wheeler N & Grice D (2000) *Management in Health Care*. Cheltenham: Stanley Thorns.

Managing at the Front Line: Managing and leading

Key points

- Effectiveness in both management and leadership are essential in social care services.
- Management focuses on day-to-day activities, whereas leadership focuses on longer-term activities concerned with transforming an organisation.
- Leadership can be shared with a wide range of staff and improves the overall performance of a service.
- Understanding the influence of emotions, both in personal terms and in relationships with others can be significant.
- Management and leadership are increasingly seen as complementary and integrated rather than separate activities.

Key words

Management; leadership; distributed leadership; emotional intelligence; transactional; transformational.

Introduction

Social care services are changing at an increasingly rapid rate. Demands of the government's modernisation agenda and the need for more control on the part of service users make it essential that frontline managers are skilled in both the management and leadership of their service. In a relatively short timescale the style of management in social care organisations has changed to encompass many more aspects of leadership, as services move from conditions of relative stability to one of constant change.

A cursory glance at the advertisements for frontline management posts in social care services confirms the wish to appoint people who are expected to 'lead developments', 'demonstrate leadership qualities', 'lead their teams' and 'lead the transformation of the service'. These advertisements suggest that organisations have an expectation that leadership is an essential aspect of managing at all levels in the service, not just at the senior levels of an organisation. Management is no longer about implementing policies decided by others remote from the front line, rather it demands that managers are now expected to lead and develop their service at whatever level they are appointed. In response to this agenda, social care services like all organisations today, place considerable faith in recruiting managers who have, or can develop, leadership qualities.

What do we mean by management?

Management is concerned with the efficient operation of a service and the development of procedures and practices that ensure that the work of the team is effective in meeting the overall goals of the organisation. When you first move into a frontline management role you will inevitably focus on ensuring that you manage the day-to-day demands on the service, and that you support your team to ensure that they are able first and foremost to provide a service that meets the needs of service users, and that also meets the strategic aims of the organisation. Your focus is likely to be on what is called the *operational management,* which consists of the short-term, routine activities concerned with the direct provision of services, as opposed to the longer-term strategic activities concerned with the future direction of the service.

Although strategy may have a long-term focus it is nevertheless dependent on the activities that take place at the front line or operational level for its success. This is where your priorities need to lie when taking on the management responsibility for a team of staff providing social care services. You are acting as a bridge between the senior and middle management of a service and the frontline activities. Strategies can succeed or fail on successful implementation.

What managers do

Whenever frontline managers in social care are asked to describe their work there is a set of commonly agreed tasks that they need to pay attention to. There can be differences of emphasis depending on organisational priorities and context, for instance:

- agreeing the plans and objectives for their part of the service
- the allocation of work and staff to cover the essential tasks
- meeting with individual members of staff for supervision, appraisal, briefing, problem-solving, etc.
- supporting and motivating staff to ensure that they perform well and get satisfaction from their work
- improving working methods to ensure that the service operates efficiently
- delegating tasks to staff to enable the manager to focus on their priorities
- leading team meetings
- supporting the development of staff
- managing the performance of the team and individuals to meet national and local performance standards
- acting as a conduit for communicating between senior management and the front line, including other organisations whose functions impact on the team

- managing change and developing a positive culture
- responsibility for controlling a budget for their part of the service
- multi-professional collaboration.

In addition, managers are also likely to be involved in:

- recruiting staff, along with others such human resources professionals or middle managers
- participating in strategic planning meetings representing local issues and ensuring that senior managers are sensitive to those issues when planning services
- representing the team and service externally
- managing partnerships with other internal departments and external agencies.

Many of the activities focus on achieving current objectives, resolving day-to-day problems and ensuring that staff meet their contractual obligations. The focus is on the present and ensuring that services are delivered effectively and efficiently in accordance with an organisation's policies and procedures. These activities can be termed as *transactional* in that they describe the relationship between manager and staff, with the emphasis on agreement about organisational goals and objectives, the work to be undertaken and the outcomes required (Burns, 1978). These are the essential processes to ensure that routine work is carried out. From this description we can see that management is mainly concerned with planning, organising, implementing and controlling. They are the main functions of management and based on the 'contract' with mutual expectations on the part of the manager and staff, and one that requires compliance and acceptance of managerial authority, although this will differ depending on the professional status and skills of the individual team members.

Moving beyond management

Once your confidence in the management role grows and you have the opportunity to reflect on your progress, you will begin to think about how you could improve the performance of your team and will find yourself becoming increasingly strategic in your thinking and actions. This means extending your concept of management and beginning to provide leadership to your team.

It is difficult to separate management and leadership. Both are needed in order that organisations function effectively and meet the needs of service users. In your first appointment to a frontline management role, you will necessarily want to focus on the management aspects of the role, but quickly you will be drawn into more strategic activity through work with senior managers on service development and with other agencies where service partnerships are developing. The more recent view of management and leadership acknowledges this reality and that there is considerable overlap between the two roles and it is increasingly recognised that they are integrated and complementary rather than separate (Skills for Care, 2006).

The views of social care workers

Discussion with social care staff suggested some of the positive attributes they valued in frontline managers who they saw demonstrating leadership qualities:

- *'a clear vision of where the team is going'*
- *'a good communicator'*
- *'an ideas person'*
- *'enthusiastic and motivated'*
- *'approachable and a good listener'*
- *'works alongside people and acknowledges their contribution'*
- *'doesn't take the credit for others' work or ideas'*
- *'supportive and takes an interest in team members'*
- *'values team members contribution and shows it'*
- *'delegates work and trusts colleagues to take decisions'.*

Some of these attributes could be considered to be qualities found in any effective manager, whereas the social care workers emphasised vision, suggesting ideas and being an enthusiastic communicator, that set some frontline managers apart and suggested they also had leadership qualities. This list resonates neatly with some of the ideas about effective leaders and suggests the style of leadership you should be developing.

Attributes of effective leadership

Wheatley and Smith (1995) have listed the key attributes that various management writers associate with leadership, including:

- an attitude to constant learning
- the development of confidence and high self-esteem
- a willingness to question and listen to answers
- a capacity for building relationships
- an appreciation of other people's work
- an ability to develop leadership in others
- innovation and initiative
- the capacity to develop a vision of the future
- an ability to communicate well at every level
- integrity and trustworthiness
- the capacity to trust others
- coaching and counselling skills.

Although Wheatley and Smith caution that this may appear like a 'wish list', these attributes correlate closely with what social care workers and frontline managers say about their colleagues who demonstrate leadership qualities. What is interesting about this list is its similarity to the list created by social care workers. It seems that effective leadership centres on a small number of personal qualities that are recognised and valued by staff. Of course, there are still areas of management that require a transactional approach, creating services and systems, but the growing demands on services and their managers suggest a much more dynamic process that creates a positive culture and takes services forward (Gilbert, 2005).

Changing perceptions of leadership

A number of radical changes in organisations have led to the development of concepts of leadership that take us beyond the established ideas based on transactional and transformational approaches and explore other models (Burns, 1978). Leadership no longer stops with managers. Rapidly changing social care environments, technologies that result in greater information dissemination and exchange, more inter-organisational relationships and networking and greater empowerment of staff, mean that many more team members are involved in decision-making and strategic activities (Fulop *et al*, 2004). These developments accord with the idea of drawing on the talents and commitment of all team members and lead naturally to the concept of *distributed leadership*, where leadership is distributed and encouraged among a wide range of staff. Leadership is not about position alone, but willingness and responsibility for all members of a team to lead. It reinforces the view that it is not only managers who can be leaders, but leadership should be encouraged from others (Seden & Reynolds, 2003) as it is important at all levels of an organisation. As confidence in the leadership role grows it is important to empower staff and encourage those with flair who can take the service forward.

> *'In developing my own style I knew what sort of leader I wanted to be. I based it on two of my previous managers who I admired. They were clear and consistent, you knew where you stood with them, they were available and supportive and thought things through in a professional way. You felt safe with them.'*
>
> Team leader, community services

Emotional intelligence

A further challenge to the traditional view that the functional skills of management are sufficient is the emergence of the concept of 'emotional intelligence' Goleman (1996). This concept is essentially about the self, which encompasses the capacity to recognise one's own feelings and those of others, and for managing one's own emotions. It focuses on how managers and leaders handle themselves and their relationships with colleagues to good effect. It is about all those elements that go to make up the individual, who is a leader and who without having self-knowledge, the skills of management and leadership would be ineffective. In working with staff you need to develop your skills in managing relationships and need to understand your own emotions, be able to manage them and recognise them in others. To be effective you need to be prepared to learn about yourself (your strengths and limitations) and be willing to make the necessary changes, although it is recognised how demanding this personal learning can be (Hill, 2003). Many of the elements that make up the components of emotional intelligence echo the kind of attributes that social care workers sought in those that led their service, which combined the functional management skills with the well-developed 'people skills' found in effective leaders.

The transformational leader

As staff increasingly expect to be 'led' by managers who share power and responsibility, and also expect to gain the collaboration and commitment of staff through shared commitment to the values and the purpose of the service, a different form of leadership is required.

Transformational leadership, a concept originally developed by Burns (1978) is increasingly common currency in many public services with its emphasis on creating the conditions for commitment to fundamental change in services, based on vision and innovation. Transformational leadership is essentially concerned with the way that leaders involve their staff in transforming themselves and the way they work. Emphasis is on inspiring staff around shared values and developing a purposeful approach to the development of the service.

Transformational leadership

- Creates a long-term vision
- Sets the broad purpose and direction
- Motivates and inspires staff by satisfying the need for achievement, recognition and self-esteem
- Develops more efficient systems
- Looks to the future and the creation of significant changes

Vision at the front line

The word 'vision' occurs frequently in association with leadership and is important to understand at all levels of a service. If you are a senior manager tasked with crafting the strategy for your organisation then vision is about the future direction of the service and its overall effectiveness. As a frontline manager, your vision is more modest, but no less valuable, but here you are focusing on supporting your staff to work well, to feel positive about their contribution and how they can improve the service, emphasising learning and development (Gillen, 2002).

Rogers and Reynolds (2003) also discuss the importance of creating and implementing a vision that they describe as '*a picture of the results you want to create, an ideal sense of what is possible and a statement of destination*' (p71). Although the frontline manager is concerned primarily with the day-to-day demands of the service and the practice of staff, they are increasingly involved in strategic planning as a result of their knowledge of need and what staff are dealing with in their practice.

> *'Part of my job is working with the senior management team in thinking about where the service is going. My particular part is concerned with employment and is expanding fast as more parents are looking for services as their sons and daughters have to move on from special schools. We are keen to develop the service so that it meets the hopes of service users and their families in line with the aspirations of Valuing People.'*
>
> Manager, learning difficulties service

These comments fit well with the arguments of Henderson and Seden (2003) when they state that the manager may not be formulating the vision, but they are designing pathways for its implementation. Acknowledging the tensions between the demands of day-to-day operations and engagement with strategic management, they nevertheless see it as giving frontline managers an opportunity to shape services and inform senior managers about local conditions and the practice environment. It is the detailed understanding of the community, the potential of the staff team and the knowledge of the practice base that can help in making a vision reality. So many strategic visions fail at the point of implementation because there has been insufficient consideration of the detailed work needed to translate vision into practice. This is where frontline managers' contribution to strategy is vital and an important and growing leadership role of the front line.

Manager or leader?

Often the terms 'manager' and 'leader' are used interchangeably, with the term leadership increasingly favoured in organisations, although there are differences at least in terms of the emphasis. Management is essentially about responsibility for the work of others and what they actually do operationally, whereas leadership is concerned with the ability to influence others towards the achievement of particular organisational goals and can be seen in a vision for a particular service.

As a newly appointed frontline manager you will be expected to manage a team of staff ensuring that the day-to-day operational activities of the organisation are undertaken efficiently. At the same time you will be expected to lead your team in creating the conditions where the service continually moves forward and changes in response to both internal and external demands and meets the wider strategic goals of the organisation. In this way you are expected to be both a manager and a leader, with one aspect of your role focusing on day-to-day performance, and the other focusing on the longer-term strategic direction of the service. Increasingly both aspects of management are now seen as important. It is no longer seen as effective to have managers just focusing on the day to day and neglecting the longer-term direction of a service. Both are seen as essential if services are to be responsive to both current needs and adapting to changing circumstances.

Strong management and strong leadership

Management in times of stability and limited change require little in terms of leadership. That situation is no longer the case and the change and complexity of modern organisations need both leadership and management combined. It can be difficult to combine the attributes of management and leadership in the same person, although this is an ideal (Gilbert, 2005). Gilbert argues that to drive change forward, leaders need to break away from the day-to-day cycle of organising and controlling (the work of the manager) and focus on listening, empowering and inspiring people. He believes that leaders need to stand back and fix their eyes on the horizon, whilst at the same time being in touch with the views of those they serve. In this view, leadership is twinned with management so that the leader is both strategic in their vision, but also attentive to the day-to-day demands on the service.

It is probably more helpful to view leadership and management in terms of distinct functions or different roles rather than requiring different types of people (Rogers & Reynolds, 2003). Mintzberg's (1990) description of the manager is that of one performing leadership roles such as being a figurehead, leader and liaison person, which sit alongside other roles that are more managerial such as monitoring, resource allocation and negotiator. To be an effective frontline manager means combining elements of both leadership and management as the service demands, recognising that it is possible to develop both roles.

The complementary nature of leadership and management is recognised in the Skills for Care (2006) report identifying what this means for social care services. They see leadership and management as integrated so that leadership is reflected in management roles at all levels. In this view, leadership and management are seen as a continuum, with the view that most leaders also need management skills and most managers are more effective if they are able to develop leadership behaviours and skills. The complementary nature of leadership and management is described in Figure 3, which illustrates the activities of leaders and managers, and also elements that are common to both roles.

A further contrast with the past is the view that managers can develop their leadership skills though training and development, and that leadership is no longer a quality that resides in particular individuals. If you feel you are not good at leading, focus on developing this part of your skill set, through developmental opportunities and study, and observe leaders who you admire. Mill (2000) suggests that as you become more confident in the management role you will begin to observe how other managers behave and what they do to motivate their staff to follow them. Also, reflect on those managers who were good leaders and what it was that made them effective in their leadership role. Mill has a number of suggestions for developing your leadership potential:

- provide your staff with a clear direction and keep the team focused on getting things done
- be enthusiastic and positive, and inspire and motivate your team
- lead by example, be honest and demonstrate integrity
- honour your commitments and keep your promises

Figure 3 The relationship between leadership and management

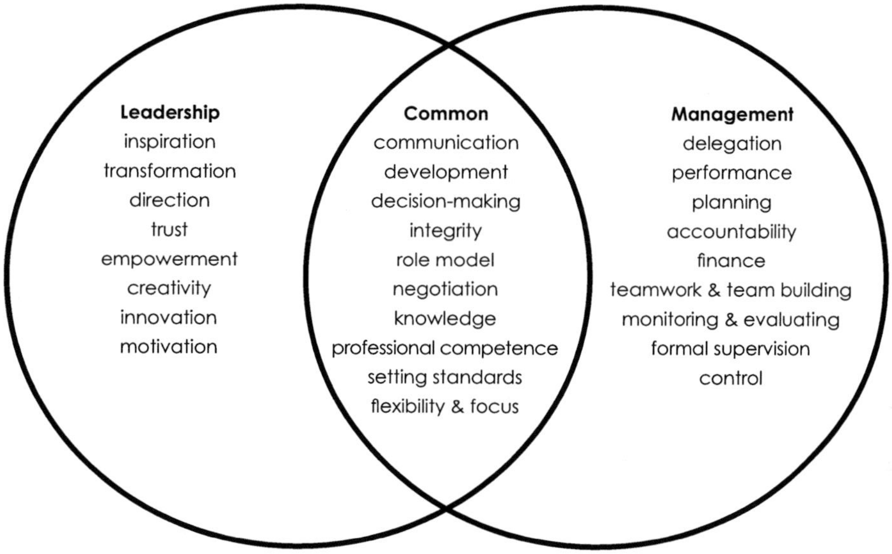

Skills for Care (2006) What leaders and managers do. Reproduced with permission.

- be a facilitative leader through coaching staff and help them to achieve their goals
- be approachable and supportive.
  (Mill, 2000: 72–74)

Mill points out that some of these skills are about you as a person, but many of them are behaviours that can be learnt over time, which you will pick up as you grow in experience and confidence in your management role. You will need to practice these skills over and over again until they become part of your management repertoire and second nature.

Conclusions

Social care organisations appoint managers who have the ability to both manage and lead, with these attributes increasingly recognised to be complementary activities. The views of social care staff suggest that those attributes associated with leadership are highly valued. Changing organisational structures and having better trained and empowered staff is leading to developments in distributing leadership more widely. These developments are mirrored by the recognition that emotional awareness can play an important role in more effective leadership. Managers wishing to develop their leadership skills should use the opportunities provided by developmental programmes and further develop through observation and modelling based on successful leaders.

References

Burns JM (1978) *Leadership.* London: Harper Row.

Fulop L, Linstead S & Dunford R (2004) Leading and managing. In: S Linstead, L Fulop & S Lilley (Eds) *Management and Organization: A critical text*. Basingstoke: Palgrave Macmillan.

Gilbert P (2005) *Leadership. Being effective and remaining human*. Lyme Regis: Russell House Publishing.

Gillen T (2002) *Leadership Skills for Boosting Performance*. London: CIPD.

Goleman D (1996) *Emotional Intelligence*. London: Bloomsbury.

Hill LA (2003) *Becoming a Manager. How new managers master the challenges of leadership*. Boston: Harvard Business School Press.

Henderson J & Seden J (2003) What do we want from social care managers? Aspirations and reality. In: J Reynonds, J Henderson, J Seden, J Charlesworth & A Bullman (Eds) *The Managing Care Reader*. London: Routledge and Open University.

Mill C (2000) *Managing for the First Time*. London: Chartered Institute of Personnel and Development.

Mintzberg H (1990) The manager's job: folklore and fact. *Harvard Business Review* (March–April) 163–176.

Rogers A & Reynolds J (2003) Leadership and vision. In: J Seden & J Reynolds (Eds) *Managing Care in Practice*. London: Routledge and Open University.

Seden J & Reynolds J (Eds) (2003) *Managing Care in Practice*. London: Routledge and Open University.

Skills for Care (2006) *What Leaders and Managers in Social Care Do: A statement for a leadership and management development strategy for social care*. Leeds: Skills for Care. Available from: www.skillsforcare.org.uk (accessed January 2009).

Wheatley R & Smith C (1995) *Leadership*. Corby: Institute of Management.

Section 2

Management techniques

Managing at the Front Line: Delegation

Key points

- **Delegation is a management process found in all organisations and enables managers to focus on their priority tasks.**
- **Newly appointed managers can find delegation difficult and often put pressure on themselves until they learn to start delegating tasks.**
- **Delegation can offer members of staff opportunities to develop their skills and widen their area of responsibility leading to greater job satisfaction.**
- **Delegation can also take place upwards where the frontline manager can depend on their own manager to take responsibility for some tasks.**

Key words

Priorities; 'letting go'; trust; greater responsibility; developing skills.

Delegation is a fundamental management process in all organisations as the role of the manager is to plan, co-ordinate and control the efforts of others. This contrasts with a manager trying to undertake all tasks themselves, which suggests that they are not managing effectively, as managing is also about getting things done through others.

A risk for the manager who does not learn to delegate is that they become overwhelmed with tasks and end up 'firefighting' in a crisis-ridden style of management, as the range of demands that require action increase in importance or turn into crisis because they are not dealt with in time. Effective delegation will free you to devote time to your priority tasks, thereby increasing your effectiveness as a manager. It will also mean that decisions will be taken as close to the front line as possible and will increase the effectiveness of your team.

The key issue is concerned with striking a balance between control and autonomy. Too much control will make staff feel they can operate only within closely defined and routine areas of work. Staff do not feel they are trusted to think for themselves and have few opportunities to exercise initiative or responsibility. On the other hand, excessive delegation can occur where staff are left to their own resources with little guidance and support (although it would be more accurate to describe this as management abdication of responsibility). Staff can feel resentful as work is forced on to them, probably because their manager has not prioritised their own workload sufficiently, and is not available to provide the guidance and support needed, particularly when there are crises and difficult decisions have to be made (Mullins, 2005).

Delegation can be difficult for the newly appointed manager who has to 'let go' of their former practice and support others to do the work they used to do. This is an important part of becoming a manager and means that you have to come to terms with letting others do the work that you used to do. It is impossible for you to do everything, but it is also part of your responsibility to develop your staff.

Some of the benefits of delegation

Delegating work appropriately to staff has many benefits both for you as a manager and for your staff. These include:

- creating more time for you to meet your managerial priorities
- allowing team members to gain specialist knowledge and skills through taking on responsibility for particular tasks
- staff becoming more motivated if they are able to gain more experience and develop new skills
- allocating work fairly among team members through the process of delegation
- decisions being made at the lowest level in the organisation that is compatible with efficiency and is most cost effective
- job satisfaction is closely related to opportunities for greater responsibility and creativity.

In discussion with frontline managers from group care settings, who were all relatively new in the management role, they described some of the reasons why they found delegation difficult:

- *'I'm new in the management role and don't know my staff well enough yet to risk delegating tasks to them'*
- *'I know I can do the job better myself'*
- *'It's quicker for me to do it'*
- *'I find it hard to ask staff who I feel are already overloaded'*
- *'Some older staff resent me asking them to take on tasks'*
- *'I'm not confident enough in my management role yet to delegate'*
- *'I'm not sure of the right way to approach staff when I want to delegate tasks to them'.*

The principle of delegation

Delegation is founded on a number of principles.

- **Authority** – the right to make decisions and take action.
- **Responsibility** – the obligation to perform a task.
- **Accountability** – which cannot be delegated as the manager is responsible for the work of members of staff. This means that the manager is responsible for ensuring that tasks are completed even when they delegate to a member of staff the actual business of doing the job. This accountability is central to the manager's responsibility (Mullins, 2005).

We can see from these principles that delegation is not just about giving staff work to do; it is also about creating a relationship between the manager and the staff member and is based on the concept of authority, responsibility and accountability. You give a staff member the responsibility for a task and the authority to carry it out, whilst you as manager remain ultimately accountable for the actions of the staff member.

Learning how to delegate is one of the most difficult areas for new managers. It is essential, and managers need to know what are reasonable demands to make on members of their staff team. The newly appointed manager who feels confident about practice issues and enjoys engaging with staff about practice, very quickly comes to learn that developing the skills of delegation is essential. It is difficult for many managers to move on from the focus on skills they used in the practice role, which was user focused to concentrate on the management of the service and their prime responsibility to achieve results through the work of their staff team.

Questions to ask yourself

When you are contemplating delegation, do you step back because of the following reasons:

- I like to appear busy or staff will feel I'm not pulling my weight
- I still like to get involved in the job from which I was promoted because that's what I know best
- I want to have a detailed knowledge of the service
- I like doing some of my activities even though they are trivial.

Issues of trust and control

Trust and control have been described as the main issues of delegation by Handy (1999). The trust is the trust that your staff have in you as their manager, and control is the control that you have over the work of staff. Handy believes that true delegation exists where there is delegation with trust and only the minimum of control. Achieving this requires a number of things:

- you must have confidence in your staff, and believe that they can undertake the task you wish to delegate
- trust needs to be given to a member of staff, and you will have to wait and see whether your decision was warranted
- trust is reciprocal, you have to trust the member of staff and they have to trust you.

As Handy argues, the proper amount of delegation and the proper blend of trust and control will depend on the situation – the degree of risk in the task delegated, the skill and experience of the member of staff, the self-confidence of the manager and the pressure they feel from their manager or the wider organisation. All these are factors that the manager must consider when contemplating delegating tasks.

The extent to which you feel you are able to delegate work to members of your team will depend on these factors, but evidence from many studies suggests that managers often find delegation a difficult area of practice. You can develop your delegation skills by reflecting on your workload and whether you have created the correct balance of tasks, your concern for control and an understanding of the skills of your staff team and wider organisational pressures.

Delegation is essentially about staff trusting each other and having confidence that work will be done, with the manager relying on staff to complete their work appropriately (Weightman, 2004). In a wider sense Weightman sees delegation as saying something about the nature of an organisation's culture and the relationship between the frontline manager and the team, with delegation an essential part of how managers influence and persuade members of staff at work.

There are some useful starting points if you are doubtful about your delegation skills, suggested by Harris and Kelly (1992).

- Look back at your time management plans. Do they suggest priority areas that you need to focus on that could be improved by delegating more tasks to members of your team? Are you dealing with areas of work that stop you focusing on your priorities?
- Are you clear about the knowledge and skills of your team members that could be utilised more fully in delegated tasks? Are there areas of work where members of staff perform well, or areas of work that could be done by staff following training?

Harris and Kelly caution that delegation does not produce results overnight, and that it may initially increase your workload – you may have to take risks and support staff. You may also feel it is 'quicker to do it yourself' and that you also lose some control of aspects of the service. However, in the longer term it is a powerful means of achieving better results in the management of the service, both for you and your staff.

Beginning to delegate

When you start to delegate you first need to get to know your staff well enough so that you have an understanding of their current level of skills and experience, their commitment and motivation and their capacity to take on additional responsibilities. This knowledge of your staff members will give you confidence in delegating tasks to them.
You could also ask the following questions.

- Is there a member of staff who can do the task as well or more effectively than I can?
- Is there a member of staff who, whilst doing the task differently, can still achieve an acceptable level of performance?
- Is there a member of staff who is paid less than me who can satisfactorily complete the task and therefore lower the cost to the service?

Feeling that you do not have sufficient work to do

One of the issues that can arise for managers when they first start delegating is that they can feel what Handy (1999) has described as under-load, where the manager feels naked and unneeded as staff take on delegated tasks. What is important to understand is that this is a natural feeling and will be quickly resolved as the frontline manager begins to identify their own priorities. It is no excuse for retaining activities that staff can more effectively complete, with its development benefits, greater job satisfaction and economic arguments.

Are staff too overloaded to take on more work?

One of the factors that can cause managers to resist delegating tasks is the concern that their staff are already overloaded, and immersed in tasks that they feel are important. This can provide an opportunity for the frontline manager to help their staff review their workload to ensure that they are focusing on the most important tasks, and to stop doing things that they may feel are important but are not essential to meet the needs of service users and the organisation, and thus provide space to take on delegated tasks.

Delegation contracting

When you delegate a task to a member of staff you need to establish a 'contract' with them where you agree on the nature and scope of the task involved, the results you want to see achieved, the timescale for completion and the extent of the authority required by the member of staff to complete the task, and finally how you will check satisfactory completion. Guirdman (1995) has described how the face-to-face discussion and assignment of the task is seen as central to delegation and more than anything else the outcome of the delegated function, with three important elements.

1. **Informational**: where the manager communicates with clarity what the task is, what outcome is sought, and what standards are expected and the resources available.

2. **Decisions**: the contracting element of the discussion with the member of staff agreeing to take on the task, also the method of work and the nature and means of reporting. This is where the extent of delegated authority, accountability and responsibility are decided.

3. **Interpersonal:** this is where the member of staff can raise issues about taking on the delegated task, with the manager reinforcing positive behaviour and establishing trust.

Delegating upwards

Coulshed *et al* (2006) emphasise that delegation is not always about delegating downwards. There is also a place for delegating upwards where you ensure that your manager gives you the support you need and also in turn safeguards the organisation by making key decisions and reviewing difficult issues or areas of functioning. The Laming Report (2003) described the risks involved when senior mangers are not informed of important decisions and frontline managers and practitioners are left to make decisions beyond their experience or area of responsibility.

Effective upward delegation means establishing a relationship with your manager where they are sensitive and receptive to issues where they need to take responsibility for decisions and support you. Developing your relationship with your manager can be enhanced by drawing on the suggestions of Harris and Kelly (1992) in their ideas for 'managing your manager', which recognises the mutually dependency between you and how you can achieve a mutually effective working relationship where your manager works for you.

Conclusions

Delegation is an essential skill that all managers need to learn, although newly appointed managers can find it difficult to 'let go' of tasks. Delegation enables the frontline manager to focus on their priority tasks and at the same time offers the opportunity for staff to take on greater responsibility and develop their skills and knowledge. Appropriate delegation of tasks is seen to increase job satisfaction.

References

Coulshed V, Mullender A, Jones DN & Thompson N (2006) *Management in Social Work* (third edition). Basingstoke: Palgrave Macmillan.

Guirdman M (1995*) Interpersonal Skills at Work.* London: Prentice Hall.

Handy C (1999) *Understanding Organizations* (fourth edition). Harmondsworth: Penguin Books.

Harris J & Kelly D (1992) *Management Skills in Social Care. A handbook for social care managers.* Aldershot: Ashgate.

Laming H (2003) *The Victoria Climbié Inquiry Report*. Cm 5730. London: Stationery Office.

Mullins L (2005) *Management and Organisational Behaviour* (seventh edition). London: Prentice Hall and Financial Times.

Weightman J (2004) *Managing People*. London: Chartered Institute of Personnel and Development.

Managing at the Front Line: Staff supervision

Key points

- The provision of good-quality staff supervision plays an essential role in contributing to high-quality services, with frontline managers promoting and sustaining practice standards through their supervisory role.
- Supervision is now recognised to play an essential role in supporting and developing staff, reinforced by the requirement that all regulated social care services should have supervision arrangements in place based on national standards.
- Supervision is seen to have four main functions: managerial, developmental, supportive and mediation.
- The development of a supervision policy conveys a positive message of its importance. Similarly a contract, between frontline manager and member of staff, establishes the basis for a more effective supervision process.
- Although supervision focuses primarily on a one-to-one relationship there is also a place for group supervision.

Key words

Staff support; frontline manager as supervisor; requirements in regulated services; four functions; policy and contract; overcoming problems; group supervision.

Supervision is one of the most important means by which social care organisations can ensure that the work of frontline staff contributes to positive outcomes for service users. It is a process that is designed to support staff to establish effective relationships with service users, develop their practice and feel confident in exercising professional judgement (Skills for Care & CWDC, 2007). It also provides the setting for establishing the accountability of the worker to the organisation and, second, promoting the worker's professional development. Through these processes supervision incorporates both the management and the professional practice systems and builds a creative relationship between the two. To achieve this, frontline managers play an essential role in promoting and sustaining practice standards, which is now increasingly recognised, for example, in the findings of the Management of

Practice Expertise project, which concluded that first line managers hold a pivotal role as arbiters of practice standards, the 'keystone' of the organisation (Kearney, 2004).

For some years, particularly with the growth of managerialism in public services that emphasised more directive forms of management and the demands of legislation such as the NHS and Community Care Act (1990), supervision took a back seat in the minds of many frontline managers as support for professional skills was regarded as less important than managing organisational processes in rapidly changing social care services. This agenda, combined with an increasing number of tasks devolved to frontline managers, diverted them from this essential activity and raised questions about the role of supervision in modern management practice. The consequence of focusing on the organisation's needs rather than outcomes for service users has been a loss of supervisory expertise and the essential bridge between management and practice provided by supervision in community-based services (Kearney, 2004).

The tragic consequences of this split was highlighted in comments following the publication of the Victoria Climbié Inquiry, which noted that the child protection system *'is built on the premise that inexperienced members of staff receive regular supervision, support and back-up from their line managers'* and that *'an over-emphasis on targets can focus on achieving success in the eyes of the Department of Health rather than in the eyes of their own frontline staff, and more importantly users'* (*Community Care*, 2001: 5).

A growing commitment to supervision

The importance of supervision is increasingly being recognised through a range of initiatives, including recommendations from the Climbié Inquiry (Laming, 2003) requirements, that all regulated services for children and adults have supervisory arrangements in place, based on national minimum standards following the introduction of the Care Standards Act (2000) (Cooper, 2002), and the General Social Care Council's (GSCC) code of practice for employers, which includes a requirement that they provide regular supervision of staff to support good practice and professional development (GSCC, 2001). More recently, Skills for Care has commissioned the development of a unit of competence to 'manage effective supervision' that describes the function of supervision for managers in social care services (Skills for Care, 2006).

Morrison (2006) has described supervision as the workers' most important relationship and argues that where practitioners are working with vulnerable or risky populations, there is no more important relationship than that with their supervisor and that the quality of services to users and levels of staff morale, skills and motivation are all bound up with supervision and the degree to which managers model the attitudes and behaviours that a service wishes its frontline staff to adopt.

Definitions of supervision

A useful starting point is to examine two definitions of supervision as this gives a clearer indication of what supervision is intended to achieve:

> *'Supervision is an accountable process, which supports, assures and develops the knowledge, skills and values of the individual, group or team. The purpose is to improve the quality of their work to optimise service users' capacity to lead independent and fulfilling lives.'*
>
> Skills for Care and CWDC, 2007: 5

> *'Supervision is a process in which one worker is given responsibility to work with another worker in order to meet organisational, professional and personal objectives. These are competent, accountable performance, continuing professional development and personal support.'*
>
> Morrison, 2006

These definitions are helpful in clarifying supervision as follows:

- an activity that is located in the context of a clearly defined relationship within an organisation
- an ongoing process
- possessing a number of objectives, including organisational, professional and personal
- involving staff who have different responsibilities
- it can include other forms of supervision than just one-to-one
- recognising the interdependence of accountability, competence, professional development and personal support.
  (Adapted from Morrison, 2006)

Functions of supervision

If you are uncertain about the supervision role, it can be helpful to have a model as the basis for action. Morrison (2006) describes four main functions of supervision.

1. **Managerial function** – this is concerned primarily with the overall performance of the worker and ensuring that their practice is in line with the agency's expectations and standards. Supervision in this function will focus on the roles and responsibilities of staff, their accountability and performance, and how well they are meeting the department's demands.

2. **Development function** – this is concerned with enabling workers to assess their own skills and competence and develop these as appropriate. Supervision in this function will focus on exploring experiences that are the basis for learning, examining knowledge, values and attitudes, and identifying the workers' strengths and weaknesses in relation to practice. Feedback will be given on performance and training needs identified.

3. **Supportive function** – where a safe and trusting relationship is shared between supervisor and worker to enable the worker to identify and deal with the personal impact of his or her work. Supervision in this function will focus on the discussion of feelings, exploration of emotional issues that may be blocking effective practice and dealing with stress.

4. **Mediation function** – this function is concerned with feedback between the organisation and the individual, and from the individual to the organisation. Here, supervision focuses on communication both up and down the organisation to provide the link between the worker and the wider organisation. It will focus specifically on organisational developments and changes, and issues about resource constraints, representing or negotiating on your behalf for a member of staff.

Supervision policy

A supervision policy sets out an organisation's statement about its expectations and requirements for supervision. It represents the organisation's views and beliefs about the value of supervision. It promotes a culture of supervision within the organisation and creates an expectation in all staff of the importance of supervision in supporting frontline staff to ensure effective practice. A policy should state as a minimum the following:

- organisational responsibility for supervision and its relationship to the overall values and aims of the service
- definitions and functions of supervision
- structure and frequency of supervision sessions
- individual supervision agreements
- resources needed to support supervision
- standards, content of recording, including confidentiality and access to records
- promotion of antidiscriminatory practice
- rights and responsibilities of the supervisor and supervisee
- periodic review of supervision arrangements.

The value of a supervision policy is particularly important in promoting the importance of supervision and conveying a positive message about the service and how staff are valued and supported through the supervisory process. It can help reduce uncertainty about supervision in both supervisors and supervisees who are new to supervision as it creates an expectation that supervision is a natural part of the process of providing a high-quality service that is responsive to service users and in turn supports its staff. It also communicates the importance of staff recognising that supervision can help them to understand what is expected of them, and that they have the necessary skills, knowledge, behaviours, values and attitudes to undertake their roles effectively and are fully supported in their work and managed effectively (Skills for Care & CWDC, 2007).

Individual supervision agreement

An individual supervision agreement (or contract) helps both the supervisor and supervisee establish the correct footing for future supervision relationship and functions. This is a process that should be established at the first supervision session and establishes the basis for how the supervisor and supervisee will work together. It establishes the 'ground rules' and the rights and expectations on both sides in creating a safe and effective supervisory setting. The contract should contain the following:

- the purpose of supervision
- practical arrangements, including the frequency and length of time of sessions, and arrangements if sessions have to be cancelled or rearranged
- the venue for sessions (a private room)
- recording and confidentiality
- arrangement for agenda setting
- arrangements for the periodic review of supervision sessions.

The individual supervision agreement should be considered a 'living document' that is negotiated and modified according to the changing needs of the supervisee, for example, as the supervisee gains greater skills and experience. The agreement should ideally be reviewed annually.

Frequency of supervision

The frequency of supervision will depend on a number of factors, including whether a service is monitored by the Quality Care Commission (CQC) where there are now minimum requirements for the frequency of supervision. In addition, where a supervisee is in their probationary period, supervision should take place fortnightly. Where supervisees are working in difficult situations, such as when they are at risk, then supervision may be arranged on a more frequent basis. For those staff not covered by CSCI inspections, such as social workers, the general principle is for monthly supervision and no less than six-weekly. Part-time staff should be supervised on a pro rata basis with agency and temporary staff supervised in the same way as permanent staff (Skills for Care & CWDC, 2007).

Unplanned or ad hoc supervision

Although most discussion of supervision focuses on planned sessions, the realities of day-to-day work in services means that staff necessarily have to 'check something out' from time to time, in order to obtain advice, guidance or direction from their supervisor. Here it is the needs and circumstances that dictate that supervision takes place 'on the spot' and is part of the normal staff–supervisor relationship, and is undertaken whilst engaged in working with service users (for example in a group care setting where the supervisee and supervisor work closely together) or when preparing to deliver a service. Where relevant any discussion needs to become the focus in a subsequent one-to-one session.

Supervision records

Recording supervision is a statutory requirement in services monitored by the CQC. Records do not have to be bureaucratic, but simply a summary of the discussion and action points is usually sufficient. Where there is disagreement regarding the content this should be recorded by the supervisor. A copy should be given to the supervisee and copies should be signed and dated by both parties. Hard copies should be taken to safeguard the supervisor and supervisee in case of any future investigations and to ensure that records are not altered in any way. In the supervision policy and agreement between supervisor and supervisee, it should state who has a right to see the supervision records. Copies of supervision records should be kept in locked storage and accessible only to the supervisor and supervisee except in circumstances where others who have the right to see them request a copy (this could include senior managers and inspectors). Copies should also be available to a senior member of staff involved in investigating a disciplinary matter.

Preparing for supervision

Ideally if you are new to the supervisory role you should not undertake supervisory duties until you have received training in supervisory techniques, or at least been briefed on the role by a member of staff who has experience in the supervisory role. Of course, you may have been fortunate to have had a positive experience of supervision in the past, which has provided you with a role model that you can use in your own practice, but more often experience of supervision is variable as a result of the lack of training, support and recognition given to supervision by many organisations.

Similarly supervisees should be briefed on the purpose of supervision so that their first experience is one where they are prepared to enter into the supervisory relationship with a clear understanding of its purpose and function, and so that it can help them to practice more effectively and support them when they are under pressure.

This is where the individual supervision agreement is valuable in setting out the purpose of supervision and its processes in a first meeting between supervisor and supervisee. It is important to establish a positive relationship from the outset as this is the key to effective supervision. You should also encourage the supervisee to view supervision as a collaborative activity where there is an expectation that they will use the supervision sessions productively.

Some problems associated with supervision

We have seen that the importance of supervision has been threatened by managerial models that undermined the importance of supporting practice, and more recently by the rapid expansion of independent sector services where, in many organisations, there was no tradition of supervision or an appreciation of its importance in safeguarding practice. Supervision, where it was provided, tended to be viewed narrowly as checking that staff are complying with procedural requirements.

This situation is beginning to change with the requirements mentioned earlier that have placed new responsibilities on services to establish supervision arrangements. However, exhortations are one thing, practice is another and my experience of working with frontline staff and their managers suggests that there is a widespread lack of understanding of the purposes and functions of supervision. Some of the main issues reflect those identified by Morrison (2006) who has listed some of the typical problems, including:

- a lack of agency policies to underpin practice
- a lack of time and space to conduct sessions
- interruptions to sessions
- supervision that is reactive rather than planned
- poor recording of supervision sessions
- focus on workload control and current crises.

The view of staff working in group care services

Discussions with staff working in group care services confirms that many of the problems identified by Morrison are familiar to this group of staff. They suggested a number of changes they would like to see that would improve the quality of supervision:

- *'supervision conducted in a relaxed and friendly atmosphere'*
- *'a better balance between service user and management issues'*
- *'maintaining regular booked supervision sessions and only cancelling for genuine reasons'*
- *'sessions should not be interrupted by telephone calls and other callers'*
- *'there should be a contractual agreement on the purposes of supervision and how it can help staff improve their practice'*
- *'supervisors to put items on the supervision agenda and not leave it to the member of staff to develop the agenda each time'*
- *'supervisors should give more feedback on practice'*
- *'supervisors should be willing to give advice and demonstrate a real commitment to practice'*
- *'there should be a focus on outcomes for the user rather than managerial monitoring of performance in a narrow sense'*
- *'there should be a willingness to examine the complex and demanding aspects of practice, particularly those which engender stress, and discussion of strategies for managing this'.*

Many of the problems identified can be overcome with appropriate training and support, allied with supervision policies that establish a basis for supervision arrangements. Much poor practice arises from ignorance of the purpose and functions of supervision. If you are new to supervision or managing or working in a service without a tradition of supervision, you will find Morrison's (2006) practical manual on *Staff Supervision in Social Care* invaluable as it comprehensively sets out the essential elements of policy and practice within an accessible format. The related manual by Knapman and Morrison (1999), *Making the Most of Supervision in Health and Social Care*, is written mainly for supervisees and will help them to understand the supervision process and how they can maximise the benefits of supervision.

Group supervision

Although the main emphasis in supervision is the traditional model of the one-to-one relationship between manager and member of their team, it is important to consider the value of group supervision. Group supervision has many elements in common with team development. The emphasis in group supervision is likely to focus on direct work with service users and carers and involve exploration of interpersonal skills, judgements and approaches used by team members (Skills for Care & CWDC, 2007). Group supervision is seen to have many potential benefits, including:

- where the numbers of team members make it difficult to provide individual supervision and group supervision is undertaken for reasons of economy
- each member learns from other people's experiences – it can also be beneficial in changing the dynamic between supervisor and supervisees and to enable other methods (eg. structured exercises, role play, sculpting and case analysis) to be used in place of the usual one-to-one relationship
- it is valuable in capturing collective expertise and collaborative effort
- it offers the opportunity to make use of a wider set of learning experiences
- it provides an opportunity for members to share their experiences and compare and contrast their own experience with the practice of others
- it can help foster team cohesion and identity.

Whilst it is recognised that group supervision can play a valuable role, as supplementary to individual supervision, it is recognised that it requires frontline managers to develop confidence and trust in working with staff teams on practice issues, which ideally grows through successfully managing team meetings, although here the focus will be different from supervision meetings.

Although team meetings have a predominantly managerial function, they can of course be reframed when necessary and used as a context for group supervision where the emphasis shifts to one of reflection on practice or case review, using the collective experience of the group to examine issues that have value for all staff.

If you want to explore the potential of group supervision then you should read Brown and Bourne's (1996) book, particularly the chapter on group supervision, which explores the issues in greater depth and poses the advantages and disadvantages of group supervision.

Don't forget your own supervisory needs

It is important that when you take on supervisory responsibilities for members of your team that you do not neglect your own needs. The reality is that many services make the assumption that once you are appointed to a frontline management post you no longer need support and supervision. As Brown and Bourne (1996) make clear, the need for supervision can be seen as some sort of weakness, a view that has been associated with the directive managerialist ethos of some organisations. They go on to argue that the opposite is true; that it is a sign of strength and professional maturity to recognise your own needs and limitations and judge when you need the support of others, particularly in view of the fact that frontline management is one of the most demanding and crucial posts.

We have already discussed the importance of your relationship with your line manager (Chapter 3) and touched on the need to express the expectation that you seek regular supervision and negotiate arrangements as part of your job role. Depending on the skills and experience of your line manager, supervision may be limited to checking accountability regarding the performance of your team, whereas ideally you will want to engage in a wider developmental relationship. If this is not forthcoming you may have to negotiate that you use other sources of support.

Conclusions

Supervision is recognised as an essential element in the management of a service. The provision of high-quality supervision plays an important role in supporting staff who work in situations where they are faced with distress or risk, and who can benefit from protected time to focus on their needs. Guidance on the content and processes of supervision is now available from a range of sources and provides frontline managers with information and developmental materials to enhance their supervisory skills. The promotion of a supervision policy and practice, and supporting a culture of supervision can help overcome some of the resistance and practical difficulties found in some social care settings.

References

Brown A & Bourne I (1996) *The Social Work Supervisor. Supervision in community, day and residential settings*. Buckingham: Open University Press.

Community Care (2001) 'Public concerns', 29 November – 5 December, p5.

Cooper J (2002) *The Care Homes Legal Handbook*. London: Jessica Kingsley Publishers.

General Social Care Council (2001) *Codes of Practice for Employers of Social Care Workers*. London: General Social Care Council.

Kearney P (2004) First line managers, The mediators of standards and the quality of practice. In: D Statham. *Managing Frontline Practice in Social Care*. London: Jessica Kingsley Publishers.

Knapman J & Morrison T (1999) *Making the Most of Supervision in Health and Social Care. A self-development manual for supervisees*. Brighton: Pavilion Publishing.

Laming H (2003) *The Victoria Climbié Inquiry*. London: The Stationery Office.

Morrison T (2006) *Staff Supervision in Social Care* (third edition). Brighton: Pavilion Publishing.

Skills for Care (2006) *Manage Effective Supervision: Leadership and management strategy*. Available from www.skillsforcare.org.uk

Skills for Care & Children's Workforce Development Council (2007) *Providing Effective Supervision. A workforce development tool, including a unit of competence and supporting guidance*. Leeds: Skills for Care.

Section 2

Chapter 7

Managing at the Front Line: Improving motivation and job satisfaction

Key points

- The policies and practices of an organisation and skilled management can make a significant difference to staff motivation and job satisfaction.
- Individual members of staff can have a complex relationship with their work and their motivation can change over time.
- The psychological contract plays an important role in the expectations of both staff and managers and needs to be balanced if both parties are to have their needs met.
- There are a wide range of management and organisational processes that can be used to increase job satisfaction.
- Although there are some broad general principles that can be applied to motivation, individual factors also play a significant part.

Key words

Changing needs; mutual expectations; sensitive management; individual needs.

Working in social care services can be physically and emotionally demanding – staff often need high levels of energy and commitment to provide high-quality services. Because staff need strong leadership and support, frontline managers play a critical role in creating the conditions where staff are motivated and experience satisfaction in their work. Where staff feel positive about their relationship with their frontline manager, they are more likely to have high levels of job satisfaction, commitment and loyalty, which in turn is associated with higher levels of performance and the willingness to make that extra contribution to the service's success (Hutchinson & Purcell, 2003). The key factor here is the strong association between frontline managers' people management skills and how staff view their employment and their job satisfaction.

Research undertaken by the Chartered Institute of Personnel and Development (CIPD, 2004) identified a range of factors that are most frequently reported in job satisfaction surveys. They include:

- having job security
- having job responsibility, and a job that stretches and challenges
- good relationships with colleagues
- a considerate manager
- good conditions of service
- fair remuneration
- good working conditions
- freedom to organise own work
- recognition for achievement, particularly feedback from the manager
- opportunities for career advancement.

This survey by the CIPD supports Herzberg's (1987) theory that factors that are considered the 'context of work' such as salary, job security, quality of supervision, corporate policy and administration and interpersonal relations all influence employees' attitude to their work. These factors, when present, provide a good working environment or work context. A further set of factors were considered 'job content', such as a sense of achievement, recognition, increased responsibility, growth and development and these are seen as motivating.

What should be remembered about all such theories is that the research tends to be undertaken with professionals and those with strong growth needs, whereas not all staff will seek more challenging work or greater responsibility. Some staff may prefer to undertake routine work with a higher level of direction and managerial control. It is important to recognise that it is not a sign of individual weakness if an individual member of staff declines to take on more responsibility (assuming they are fulfilling their contract obligations). Individuals have complex orientations to their work and the challenge for managers is to better understand their staff so that they can match the demands of the job with the expectations of staff (Fulop & Linstead, 2004).

It is also important to remember that a member of staff's attitudes to their work changes over time as their personal circumstances change. This can mean that someone who has not been able or willing to take on more responsibility can, as a result of other changes in their life, or new opportunities or incentives, wish to develop their career or skills. The reverse situation is also possible, where new personal demands result in a reduced commitment to work responsibilities. The message for the frontline manager is that people change and their attitudes and expectations about their work also change over time.

What frontline managers can do to improve motivation?

There are a group of well-established strategies (Mullins, 2005) that frontline managers can use to improve motivation and that are seen to lead to greater job satisfaction.

- Job enrichment – where steps are taken to make a job more interesting by providing staff with opportunities to take greater responsibility, autonomy and control over the work they perform. This is seen to stretch the individual and provide a greater sense of achievement.
- Situational management – which recognises that no one form of management is appropriate in every situation and that it is dependent on the situation, the task and the individual. Where frontline managers can adapt their style of managing to take account of the needs of members of staff, it is more likely to motivate the individual. In other words, motivation and management style are seen to go hand in hand.
- Participation – tends to result in increased commitment from individuals, with the feeling of ownership in processes. Most staff value being involved in discussions where they can contribute to the quality of decisions made. Of course, participation must be genuine. Where decisions are taken irrespective of a participant's involvement then managers loose credibility.
- Delegation – is seen to lead to improved morale by increasing motivation and job satisfaction as it helps to satisfy an individual's higher level achievement needs. It can create conditions where staff are able to plan how their work is to be done, which in turn is seen to encourage positive attitudes to their work.
- Flexible working – where managers are able to respond to the needs of staff and the demands made on them, particularly in creating a healthy work/life balance. This is most often seen in flexi-time arrangements, job sharing and staggered hours to meet family and other demands.

Managers also need skills in overcoming barriers that stop staff performing well, and which can include political, bureaucratic or resource barriers. Where managers use their power and influence to change situations this is valued by staff and contributes to them feeling that their manager is sensitive to their needs and is working for them in tackling problems that need management authority.

The importance of the psychological contract

All employees of social care organisations have a legal contract with their employers, which states what each party is expected to deliver. It is recognised that a legal contract of employment offers a limited representation of the actual reality of the job and the employment relationship. Whereas the psychological contract is an unwritten and often non-verbalised set of expectations and assumptions about the obligations that people ascribe to their organisation (Linstead *et al*, 2004) – it is the perceptions of the employee and employer of what their mutual obligations are to each other. These obligations will often be informal and imprecise and are likely to be inferred from actions as well as statements made by the employer (CIPD, 2004).

The psychological contract is essentially a set of expectations, that the member of staff expects from the organisation and in turn what the organisation expects from the member of staff. Where a contract is not perceived identically by both parties it is a potential source of conflict. For example, the member of staff may feel exploited by the organisation as it expects more than they can provide within their contract and on the other hand the organisation feels that the individual is uncooperative (Handy, 1999).

In a workshop with frontline managers and staff from group care services, the following were seen as some of the important areas of the psychological contract.

Care staff expectations of their organisation

- *'It provides job security'*
- *'It provides challenging and satisfying work'*
- *'It consults with staff and allows genuine participation in decisions which affect the individual'*
- *'It rewards staff fairly'*
- *'It recognises effort and achievement'*
- *'It provides opportunities and funding for training and development'*
- *'It treats members of staff with respect'*

Managers' expectations of staff

- *'They adhere to the policies and procedures of the organisation'*
- *'They respond to the reasonable authority of managers'*
- *'They demonstrate loyalty and commitment to the organisation'*
- *'They are punctual for work and do not take unauthorised absences'*
- *'They are willing to change their practice when it is in the interests of service users and the wider service'*
- *'They are willing to work co-operatively with colleagues'*

Value of discussing the relationship between the individual and the organisation

Because psychological contracts are largely unspoken and therefore not defined formally, they can affect the relationship between employee and employer (typically the frontline manager and members of the staff team). It is important to explore expectations in order that there is a better 'fit' between the expectations of the employer and employee, as an unbalanced contract will influence the individual employee's attitude and commitment to the organisation. The clearer that both parties are about their expectations, the closer the fit and the likelihood that they will have their expectations met.

The CIPD (2004) report sees the psychological contract as having clear implications for the way that frontline managers act in relation to their staff.

- Process fairness – staff want to know that the organisation takes an interest in them and will take their needs into account when important decisions are taken that affect them.

- Communication – a two way dialogue between employee and employer is necessary as a means of giving a voice to the employee. Informing and consulting employees are more likely to see outcomes as fair where they are involved.
- Managing expectations – managers need to be frank about what newly recruited staff can expect from the job. The difference between rhetoric and reality is quickly spotted by employees and can undermine trust in managers. The recruitment process can be used to clarify the 'deal' between employer and employee.
- Measuring employee attitudes – monitoring employees' attitudes to their work and the organisation on a regular basis is one means of identifying employee satisfaction. Employees do not often share senior managers' view of reality. Where surveys are undertaken employers should only use surveys if they are willing to act on the results.

Some of the key messages from the CIPD report suggest that where frontline managers have good people skills, employees are more likely to trust their line manager. Frontline managers must ensure that their staff have the opportunity for more responsibility as autonomy increases satisfaction, and tight management and supervision tends to reduce employee satisfaction. If managers can't keep a promise they need to explain why, as failure to do this can lead to a loss of trust. Finally, managers should not rely on performance management to motivate employees, but engage them with hearts and minds.

Job satisfaction

Job satisfaction is an all-embracing concept that focuses on the attitudes that members of staff have towards such things as salaries and wages, human resource policies, the style of their manager and relations with other members of staff. Understanding these factors is important in organisations as they are seen as significant in terms of staff turnover (Hannigan, 2004).

As with motivation, there are a wide range of theories of job satisfaction, but what is clear is that there seems to be a number of factors that suggest that job satisfaction is related to a small range of key factors and that when these are present staff are likely to be more satisfied in their work. None of the factors should come as a surprise as they are all factors that are seen as important in well-managed organisations.

- The organisation supports flexible working arrangements where staff feel their manager is sensitive to their needs.
- Staff feel that they are able to achieve a satisfactory work/life balance, with the quality of working life influential in how staff feel about their employment.
- Involvement and empowerment with emphasis on sharing information with staff, pooling ideas and sharing concerns with employees and the development of mutual trust.
- Job enrichment, through greater responsibility, delegation and participation in self-managed work groups.

Job satisfaction is also related to the demands that the job makes on individuals. If a job is demanding, and staff feel they have a high level of autonomy in how they do the job, this tends to lead to much greater job satisfaction. Where a job is easy or requires a lower level of skill than that possessed by the member of staff, little learning takes place. Demanding jobs and the freedom to decide how to do the job is said to stimulate continuous learning (Hutchinson & Purcell, 2003).

The important lesson from studies of job satisfaction is the need for managers to match the job demands with staff skills and abilities. Studies have shown that where there is good matching in terms of skills, abilities and job demands, motivation, performance and job satisfaction tend to higher. Conversely, where there is a poor fit between a member of staff and the job, their commitment and job satisfaction was low (Hannigan, 2004).

Evidence from countless studies of what staff seek in their employment points to a particular range of factors that constitute the basis for high motivation and job satisfaction:

- understanding the aims of the organisation and valuing its ethos and direction
- feeling that they are valued by the organisation and particularly have a good relationship with their line manager
- having a good working relationship with colleagues
- having the opportunity for professional development
- feeling that they are able to participate in what happens in the organisation
- receiving consistent support and supervision.

We have seen from the different surveys and theories about motivation and job satisfaction that there are a range of factors that are potentially influential, and that these are important to consider when creating workplace conditions where staff are motivated and receive satisfaction from their work. Nevertheless, it is important to be cautious and recognise that there are no specific rules that work with all staff in all situations. Although there are broad principles that you should adopt in creating a positive work environment, you will need to adapt your approach to allow for the expectations, values and skills of individual members of staff, recognising that motivational factors may vary from person to person.

Conclusions

There is a strong relationship between staff who are highly motivated and experience job satisfaction and the quality of a service provided to users. A key factor in motivation is the relationship between the frontline manager and a member of staff, with a well-established series of factors that are linked with increased motivation and job satisfaction. The recognition of the psychological contract and the mutual expectations and obligations between an organisation and staff also plays an important role in creating the conditions for increased motivation. Ultimately motivation and job satisfaction are rooted in individual factors and the needs of individuals should be considered in the context of organisational policies and practices.

References

CIPD (2004) *Managing the Psychological Contract*. Factsheet. London: CIPD. Available from: www.cipd.co.uk.

Fulop L & Linstead S (2004) Motivation and meaning. In: S Linstead, L Fulop & S Lilley (Eds) *Management and Organization. A critical text*. Basingstoke: Palgrave Macmillan.

Handy C (1999) *Understanding Organizations* (fourth edition). London: Penguin Books.

Hannigan T (2004) (Ed) *Management Concepts and Practices* (fourth edition). Harlow: Pearson Education.

Herzberg (1987) One more time: how do you motivate employees. *Harvard Business Review* **46** (1) 109–131.

Hutchinson S & Purcell J (2003) *Bringing Policies to Life: The vital role of the front line managers in people management.* London: Chartered Institute of Personal and Development.

Linstead S, Fullop L & Lilley S (2004) *Management and Organization. A critical text.* Basingstoke: Palgrave Macmillan.

Mullins L (2005) *Management and Organisational Behaviour* (seventh edition). Harlow: Pearson Education.

Section 2

Chapter 8

Managing at the Front Line: Managing and developing teams

Key points

- Teams play an important function in social care services where the collective knowledge and skills of members can improve the overall quality of a service.
- There are a range of factors that influence effective teamwork and these should be understood by the frontline manager when managing and developing their teams.
- Teams experience different stages of development that influence how they function at a particular point in time.
- 'Groupthink' can develop teams as they become too cohesive and resist new ideas.
- There are practical activities that frontline managers can use to develop their teams.

Key words

Common characteristics; communication; stages of development; trust and respect; conflict; development activities.

Much of the work in social care services is undertaken by teams, with staff appointed to work together to achieve organisational objectives. This requires the collaborative efforts of staff with different skills and knowledge, working within and across organisational boundaries, this requires considerable commitment on the part of the frontline manager and team members if they are to work through the myriad issues that can confront team working.

What is a team?

First, it is important to differentiate between a group and a team. Groups could be described as a collection of people who interact with one another because they have a similar purpose or interests. A team is a group that has a common goal or task, and the achievement requires that members collaborate and co-ordinate their activities and have regular contact with each other in the pursuit of the task (Martin & Henderson, 2003).

In defining what we mean by a team there are a number of common characteristics that are shared by most work groups (Adair, 1987), including:

- a definable membership
- a group consciousness, with members having a collective perception of unity and identification with each other
- a shared purpose, with members having the same common task or goal
- interdependence, with members needing each other's support to accomplish the task
- interaction, with members communicating and influencing each other
- acting in a unitary manner, with the group working as a single entity.

A helpful definition of a team emphasising the importance of integrating the work of each team member states that *'A team is a group of individuals who work together or deliver a service for which they are mutually accountable. Team members share goals and are held mutually accountable for meeting them and are interdependent in their accomplishment, and they affect the results through their interaction with one another'* (Mohrman *et al*, 1995: 39–40).

Why are teams needed?

Teamwork is seen as essential in achieving high-quality social care services as it requires groups of staff to work together for a common purpose. This could be for longer-term mission-driven purposes or short-term project purposes. It is also seen as important where it is necessary to draw on the expertise of a range of professionals, and where there is greater task uncertainty it can represent the different perspectives of those involved (Martin & Henderson, 2003). For effective teamwork it is essential that staff communicate well with each other, and none of this happens without good planning and effort. For teams to work well, team members need support and nurturing, both as individuals, so that they feel their work is recognised and valued, and in learning to work as a member of a group and giving support to each other. It is here that the frontline manager plays a vital leadership role in both managing teams and developing their potential. The quality of teamwork will differ in different environments. For example, work in group care settings, where there is greater need for interdependence among team members as they need to rely on each other more directly, through activities such as physical caring tasks or where there are difficult incidents that need managing, than say staff in fieldwork settings where independent working are more common. The key point from these different styles of working is that staff need to communicate and trust one another in working safely and effectively (Ward, 2003).

Teamwork means all team members

Several frontline managers interviewed commented that any discussion of team management should reinforce the point that teams mean everybody in the team, including administrative and support staff. They recognised the tendency for these staff to be ignored, yet they play an important role in enabling other staff to meet the objectives of the service. As a frontline manager you need to apply the same principles to administrative and support staff as you would to social care staff.

Managing a team

In teams consisting of a range of staff with different professional backgrounds and responsibilities, it is necessary that each play their own role, but also integrate their work closely with that of their colleagues. It is important to understand that developing effective management of individual members of staff is not the same as managing a team (Hill, 2003). You should not equate managing the team with managing individuals. To feel confident in managing teams you need to know something about what an effective team is and the dynamics involved in designing teams. This requires an understanding of what constitutes an effective team and the dynamics involved in developing the team.

Acknowledge the differences in the team

Whether the team is made up of staff from one profession or a range of professions it is likely that there will be different status, values, culture and conditions of service among its members. If the team is also made up of different professions there will also be different codes of practice and priorities and you need to understand the consequences of a mix of different value bases and theoretical frameworks, the changing membership of the team and the influence of the culture of the organisation on the team (Statham, 2004).

The relationship between the leader and the team

The key to effective leadership of teams is one where the relationship between the frontline manager and the team is based on trust and support, where there is an understanding of the needs of team members and your relationship with them is the most important factor in the success of the team. As the frontline manager you have the tough job of balancing the sensitivity and support elements of team management with the need to establish clear goals and performance requirements to achieve objectives. One of the elements you need to consider to lead your team effectively includes open communication. A manager should be willing to model open communication by being frank and open about issues they wish to discuss with the team. This will encourage team members to be more open, giving members the scope to use their own judgement as much as possible and amounts to an offer of trust by the manager.

A further important factor is that individuals learn more quickly when working in teams as they better understand the shortcuts and ways of doing their job. This means that frontline managers are critical agents in creating effective team functioning where individual members of staff can learn 'on the job' (Hutchinson & Purcell, 2003).

Defining the team and its responsibilities

Having a clear view of what is expected of the team is helpful in defining the purpose of the team and being clear about your responsibilities. You (and team members) need to be clear about the team's mandate as this forms the contract for the work of the team in which both sides (managers and team members) have responsibilities and rights. This is not something that is passed down from senior managers, but is clarified through ongoing discussion and negotiation between the managers and the team. The importance of this activity is that it can help eliminate misunderstanding, fantasies and covert action about the expectations, tasks and resources of the team (Morrison *et al*, 1987).

Influence of team size

Although managers find themselves responsible for teams of widely varying size, it is argued that as teams grow in size they become more difficult to manage (Mullins, 2005). Teams of six to 11 members are considered an appropriate size. Teams of this size are easier to manage as they can avoid the risk of subgroups developing, but can become too cohesive and resist change. Where teams are larger, say over 15 members, there may be a greater range of skills and experience, but there is a tendency for less interaction and less opportunity for group participation, with the team meeting less frequently and the sharing of communication or sharing of common and agreed tasks more difficult. Larger groups risk splitting into subgroups with rivalry and reduction in cohesion. The reality of much social care is that teams can be considerably larger than those suggested as most effective by Mullins, with teams of 20 plus staff are common. The implications are that managing such teams is much more demanding on frontline managers and the reality suggests that it will not always be possible to create the cohesive team working suggested by management theorists. If this reality is accepted then it is possible to develop structures and processes that enable large teams to function effectively. It is here that concepts of distributed leadership and self-managed teams become important.

Characteristics of team members

For effective team working, individual members have to work closely together, support each other and balance the team's needs and their own. There are strengths and weaknesses in the make up of teams. Those teams that are more homogenous and consist of staff from one type of background may work together well, but are likely to view issues from one single perspective. They risk becoming too cohesive and susceptible to groupthink, which includes a propensity to drift and not examine their performance.

There may also be pressure on individuals to conform with a search for consensus and a suppression of unpopular ideas. A more heterogeneous group of staff with different skills and backgrounds may have a wider range of perspectives on issues. Here the complementary experiences, skills and styles of working can lead to greater creativity, innovation and problem-solving. It is likely that, as a newly appointed frontline manager, you will inherit an existing group of staff but, over time, you should strive to create a diversity of team members who have a range of skills and experiences. There is the risk that as the manager you will tend to appoint staff who are 'in your own image' as there is a tendency to be attracted to and most comfortable with people who are like yourself.

Building trust and respect among team members

The greater the degree of trust and respect in a team, the greater the likelihood that members will be sensitive to each other's needs, therefore they will be more productive than where members are cautious about sharing and suspicious of each other. Trust can grow as team members meet to explore their values and ideas about the service in developmental activities, and work alongside each other growing in respect for each other's contributions. Frontline managers can set the tone for the growth of trust in their leadership of the team, modelling the behaviours that can enable team members to develop trust and respect for each other, and shaping the team's culture in terms of norms, routines and values, setting the tone for team interactions that become the standard for how team members behave towards each other.

Developing teams

One of your roles as frontline manager is to lead your team and release its potential to enable it to achieve results greater than the sum of its parts. To do this successfully means being aware of those factors that influence the performance of teams and how you can improve team working through developmental activities. Teamwork involves both content and process. Group processes focus on issues such as team morale, feelings, conflict, competition and co-operation. Less emphasis tends to be placed on process even though it can be a major cause of ineffective group action.

Developing an effective team takes time and is less easy to accomplish if time is not taken away from the pressures of service delivery to reflect and explore the work of the team and how each member can contribute to achieving the goals of the team through team-building activities.

The Adair (1988) leadership model with its focus on task, individual and team can be used to understand the constituents of team effectiveness. First, the individual's needs have to be satisfied. These are the needs that the individual brings with them into the team. Second, there needs to be successful team interaction, where the group builds, develops and maintains working relationships among its membership so that they can work together effectively to accomplish tasks. Third, is the achievement of the team's task, which is a measure of the team's effectiveness. Though it should be noted that the focus on the task can risk neglecting the process of how the team works together. The reality of much teamwork is that energy is devoted to task achievement. Therefore, it is here that sensitive leadership of the team can create opportunities for the development of how the team can work together and at the same time satisfies the needs of individual members.

What constitutes an effective team?

Effective teams exhibit a range of characteristics (Mullins, 2005), including:

- commitment to shared aims and objectives
- a sense of commitment to the group
- the acceptance of group values and norms
- a feeling of mutual trust and dependency
- the full participation by all members and decision-making by consensus
- a free flow of information and communication
- an open expression of feelings and disagreements
- the resolution of conflict by the members themselves
- a lower level of staff turnover, absenteeism, accidents, errors and complaints.

In discussion with frontline managers from group care settings they identified some of the factors they sought in effective teams:

- *'good leadership'*
- *'clear understanding of the purpose of the team'*
- *'the leader briefs the team regularly on changes in the organisation that affect the team'*
- *'good working relationships within the team'*
- *'enthusiastic team members'*
- *'team members have the confidence to express their opinions'*
- *'regular team meetings'*
- *'regular team development sessions'*
- *'that the team has fun'.*

They also identified some of the factors that led to less effective teams:

- *'the manager doesn't communicate team issues to senior managers'*
- *'constant turnover of team members'*
- *'individuals who dominate meetings'*
- *'team members who have their own agendas which are not discussed in the team'*
- *'important decisions get taken outside the team meeting'.*

How groups develop

To understand how a team develops its effectiveness it is useful to examine the model of four stages of group development originally developed by Tuckman (1965), and two additional stages of conforming and mourning added by Gersick (1988).

- **Forming** – when group members find out about such matters as 'what do others expect of me?', 'how do I behave in this group?' and 'what are the rules in this group?' This is the stage where the individual is trying to establish their personal identity and make their mark on the group.

- **Conforming** – when team members fall into an easy consensus in order to feel as though they are really working together.
- **Storming** – when conflict occurs while group members sort out roles and relationships in the group. This is the stage where personal agendas are revealed and some interpersonal hostility is present. This is also the stage where the norms of the group are tested.
- **Norming** – when co-operation develops and new 'norms' are developed, including mutual support and the development of a group feeling. This stage is marked by the group establishing how it should work, how it makes decisions and the degree of openness, trust and confidence appropriate to the group.
- **Performing** – when the group is operating effectively as a team, with constructive attempts at task achievement.
- **Mourning** – sadness when teams break up, members leave or projects come to an end, with members becoming nostalgic as they try to recreate former team working in new projects, with reference to former ways of working, prolonging the reluctance to move on and learn new things.

It is important to remember that the stages in group development are not necessarily sequential, and that groups can move back and forth between different stages, particularly as group membership changes. When there are significant changes in membership of the group it can experience changes that have been described as *re-forming*. Gersick (1988) has also argued that if groups fail to experience the storming/conflict stage in the early stages this can result in differences being suppressed that can push a group into reassessing its aims and changes in direction.

Frontline managers from group care services described the difficulty in supporting the development of a mature team when faced with continual turnover of staff. Collectively they voiced the following view.

> *'We rarely seem to reach a stage of performing. We just get the team established when there's staff movement and the team has to re-form again. You have to constantly remind yourself that you are working with teams that are going through different stages of development and this has implications for how you manage them.'*

Self-directed teams

Although this is a concept that has been developed in industry, it has been adopted by some social care organisations where it is recognised that a frontline manager responsible for a large number of staff cannot practically offer the level of management support possible in smaller teams. Specific goals are set for teams, with responsibility delegated to the team who have the freedom to plan and control their own work, with the level of supervision reduced. In such teams, members agree the planning and scheduling of work, agreeing tasks within the team and collective agreement on standards of performance and corrective measures. It has value where managers have very large teams of staff who inevitably work more autonomously.

Giving teams greater responsibility for managing their own performance is said to be strongly related to increased job satisfaction (Mullins, 2005). The concept of self-directed teams shares some of the same ideas as distributed leadership where a wider range of staff take on leadership responsibility. Mullins argues that self-directed teams are still answerable to the management of the service but nevertheless it is seen as a way of ensuring that teams develop greater independence and responsibility for their own performance.

Conflict in teams

One area of teamwork mentioned by staff interviewed for this handbook was managing conflict, particularly between members of staff teams.

Different views of conflict

Conflict is usually associated with negative features. It can be viewed as an expression of feelings and is a form of behaviour that obstructs the achievement of another's goals. On the other hand, it can be seen more constructively as something that can facilitate improvements in services.

Some sources of conflict in teams

- Where there are differences in perception about important issues.
- Where the work of one member of staff has consequences for another.
- Where there are values that are deeply held.
- Where there are individuals narrowly concentrating on particular areas of practice to the neglect of other areas.

Constructive outcomes of conflict include:

- producing new ideas that can lead to solutions to problems
- encouraging team members to seek new approaches
- bringing problems to the surface
- enabling individuals to clarify their views.

Negative outcomes of conflict include:

- mistrust and suspicion developing in the team
- team members becoming defensive
- individual interests taking precedence over team objectives
- can lead to people leaving the team
- polarising opinion within the team.

Managing conflict

How you decide to manage the conflict will depend on whether you see it as negative or constructive and the extent to which it impacts on the service. Mullins (2005) has suggested some strategies that can reduce the negative effects of conflict. These include the following:

- Focusing on superordinate goals that are shared by all members of the team can help to defuse conflict and lead to more co-operative behaviour. The focus on the shared purpose of the group is described by Handy (1999) as *'drenching the organisation in a common set of values'*.
- Developing interpersonal and group process skills, which can enable team members to better understand their own behaviour, the viewpoint of others and communication and problem-solving processes.
- Working on group development activities, particularly those activities which increase group cohesion and reduce dysfunctional conflict.
- Creating a climate of group leadership, which is participative and supportive is likely to influence the level of conflict by demonstrating an attitude of respect and trust, encouraging personal development and creating an environment where staff work co-operatively together.
- Distributing resources equitably to help overcome conflict about the differences in resources within the group.

Practical activities to develop your team

Finally, in developing teams there are a range of practical activities that you can undertake that can help support the development of your team. These include:

- promoting the purpose of your team that will act as a motivator
- encouraging openness and involvement of all members in team meetings
- encouraging team members to support each other, setting the tone by your own supportive behaviour
- encouraging the team to deal with differences of opinion – being willing to compromise and come to agreement on action with all team members supporting the decisions taken
- reviewing the performance of the team periodically – identifying what worked well and where improvements are needed, with agreed action taken when performance does not meet agreed standards
- supporting your team in its decision-making functions, based on information gathered, problem recognition, discussion and evaluation of options, before they commit themselves to action
- promoting the team both within your organisation and externally, enabling the team's work to be recognised – make certain that the team's activities integrate with the wider organisation
- celebrating the success of your team through regular feedback, using praise and publicising the success of the team.

Conclusions

Teams constitute a common feature of most social care services where staff come together and work collaboratively to enhance the quality of a service provided. A range of factors, including size, relationship with the leader and other team members, and issues of trust and respect can all influence team working. Teams are continually developing and changes in membership can mean new stages of development with consequences for team effectiveness. Conflict in teams is common and can express different views about the values and purpose of the team. The resolution of conflict can enable teams to perform more effectively. There are a range of activities that frontline managers can initiate to develop the teams they manage.

References

Adair J (1987) *Effective Teambuilding*. Aldershot: Gower.

Adair J (1988) *Effective Leadership*. London: Pan Books.

Gersick C (1988) Time and transition in work teams: towards a new model of group development. *Academy of Management Journal* **31** (1) 9–41.

Handy C (1999) *Understanding Organizations.* London: Penguin Books.

Hill C (2003) *Becoming a Manager. How new managers master the challenges of leadership*. Boston MA: Harvard Business School Press.

Hutchinson S & Purcell J (2003) *Bringing Policies to Life: The vital role of front line managers in people management.* London: CIPD.

Martin V & Henderson E (2003) *Managing in Health and Social Care*. Abingdon: Routledge.

Mohrman SA, Cohen SG & Mohrman AM (1995) *Designing Team Based Organisations*. London: Jossey Bass.

Morrison T, Waters J, Roberts W, Craig E & Shearer E (1987) *Surviving in Teams.* Rochdale: NSPCC Protection Team.

Mullins L (2005) *Management and Organisational Behaviour*. Harlow: Prentice Hall and Financial Times.

Statham D (2004) The context for managing practice. In: D Statham (Ed) *Managing Front Line Practice in Social Care*. London: Jessica Kingsley Publishers.

Tuckman BW (1965) Development sequences in small groups. *Psychological Bulletin* **63** (3) 384–399.

Ward A (2003) Managing the team. In: J Seden & J Reynolds (Ed) *Managing Care in Practice*. London: Routledge and Open University.

Section 2

Chapter 9

Managing at the Front Line: Managing meetings

Key points

- **It is important to ask whether a meeting is the most suitable way of communicating information and making decisions.**
- **Meetings can be time consuming and costly and still not lead to effective outcomes.**
- **Understanding the purpose of a meeting can help you decide whether it is the most suitable way of achieving its objective.**
- **Meetings that are skilfully planned, chaired and recorded can be productive both in process and outcomes and more satisfying for participants.**

Key words

Purpose of meetings; costs of meetings; agendas; chairing; participation; record of meeting.

A considerable amount of management time is taken up chairing and attending meetings, both with other managers and with members of staff. Badly run meetings are costly in time for all involved and the true cost of holding meetings is rarely calculated. As services have become more complex with increasing cross professional and organisation processes, the time spent in meetings can often outweigh that spent actually providing a service. Although some meetings are essential, many seem to be arranged where the purpose is unclear or as part of a pre-arranged schedule even though there may be few items actually requiring discussion.

The planning and management of meetings is essential if time is to be used productively. Of all the activities that take place in social care services, meetings come in for more criticism than most other activities. Yet they are important in bringing staff and others together where joint discussion and decision-making is necessary. An understanding of how meetings can work and the skills needed to manage meetings effectively are essential for frontline managers (Coulshed & Mullender, 2006).

In this chapter we will consider the usual form of meetings that frontline managers have to chair as part of their responsibility to communicate with groups of staff, for example team meetings. Where meetings are held with large groups of staff these are more accurately described as presentations as the size of a group severely reduces two-way dialogue both with the chair and within the group.

Why do we need meetings?

The purpose of meetings is fivefold:

1. To share information with staff, particularly where the actions that follow need discussion among those present. For example, in working in a co-ordinated manner to deliver a service, for example in a team meeting.

2. To provide consultation on proposals where participants' views are important and where the implications of a proposal need joint discussion, for example when there are proposals for changes in a service.

3. To collect information prior to making decisions.

4. To make decisions, where the participants' views and knowledge on the options are important before a decision is made.

5. To provide information and guidance where group discussion is important.

When meetings are effective they:

- are necessary
- communicate important information to the staff group
- improve the quality of decision-making through group discussion
- provide a forum for structured discussion
- contribute to team development.

When meetings are ineffective they:

- appear to have no purpose
- make straightforward issues more complicated
- last too long as a result of poor timekeeping
- provide a platform for verbose individuals or those with an 'axe to grind'
- delay decisions and action.

What needs to be considered before a decision to hold a meeting is made

- Is the purpose of the meeting clear and is meeting the best way to achieve it? Avoid arranging meetings unless you can clearly state the purpose of the meeting or it is essential to meet some organisational objective. Consider other means of communicating with members of staff, through email or telephone calls.
- Who needs to attend and does everyone need to be there? Are the correct people attending the meeting who can all make a genuine contribution to the meeting and are they in a position to make decisions?
- Where will the meeting be held? Is the venue conveniently located closest to where most participants work and therefore avoids unnecessary travel time? Is the venue conducive to a productive meeting?
- When will the meeting take place? Is there a preferred time of the day?
- Is there preparation required for the meeting?
- Is the agenda planned in advance and circulated or at least posted up where staff can see what is on the agenda?
- Minutes are important as they are a record of what happened at the meeting, particularly the decisions made and who has responsibility for actions following the meeting. They are also important as a record for those staff who were unable to attend.

Discussion with frontline managers from group care services identified some of the problems they experienced in leading team meetings:

- *'I find it difficult to get people to speak out'*
- *'how do I stop a team member taking over the meeting or "rambling on"'*
- *'dealing with a member of staff who influences others'*
- *'it's difficult to get feedback from staff'*
- *'it's hard work getting staff to put items on the agenda'*
- *'how can I make sure we finish on time?'*

Chairing meetings

Managers need to develop the skills that allow them to chair meetings effectively. A well-chaired meeting is successful both in terms of achieving its purpose, but also in efficiently taking the minimum amount of time commensurate with the length of the agenda and the decisions that need to be made. As a chair you need to be firm in moving the agenda on, but also sensitive to the needs of staff who wish to make their points. You need to be able to handle interruptions, time wasting, irrelevant information and domineering behaviour without offending people. This is best done by gently reminding people of why they are meeting and what they need to focus on and the decisions to be made. Always start on time. If you delay the start because some staff are late this is disrespectful to people who attend on time and also sets a precedent for those who know they don't need to attend on time.

Chairing a meeting: the process

- Start on time.
- Welcome everyone.
- Ask for and record any apologies.
- Check that everyone has an agenda and a copy of the minutes of the last meeting, and have sufficient copies for those who have not received these or forgotten theirs.
- If you do not have someone to take minutes ask for a volunteer as the recorder. You should avoid taking the minutes yourself as this can distract you from focusing on the discussion with colleagues. Minutes do not need to be lengthy – a brief record of decisions is usually sufficient.
- Briefly review the agenda so that everyone's thoughts are focused on the purpose of the meeting.
- Ask if there are any matters arising from the last meeting and deal with these immediately, unless they are deferred as they are on the agenda for the current meeting.
- Briefly go through the agenda identifying items that are for information only, for discussion and those where decisions are needed.
- Take the items that are likely to be controversial or require more time early in the meeting and leave items that are likely to unite the meeting till later.
- Intervene tactfully if someone hogs the discussion. Try to bring quiet members into the discussion.
- Try to draw out quieter members, for example, '[name] can you tell us how you dealt with these issues last time?' or '[name] do you have any ideas about how we could improve things?'
- Try to avoid staff using meetings as the opportunity to raise a wide range of issues where there isn't time to consider them properly and individuals can leave frustrated. Encourage staff to put such items on the agenda for a future meeting. Where an individual has a hidden agenda and the meeting starts to discuss a particular issue remind participants of the need to stick to the agenda. A suggestion to convene a separate meeting to address the purpose of the hidden agenda can be an effective way of dealing with the issue.
- If a member of staff keeps criticising the service, or what is wrong, give the problem back to them. Ask them how they would deal with the situation. For example, 'I understand you are concerned about [subject] what do you think would be the best way to deal with this?'
- As you progress through the agenda summarise the main points of agreement and the decisions made as this will tend to keep the meeting moving, and it will help prevent discussion drifting back to topics already covered.
- If the meeting gets bogged down you could use the following form of words: 'Can we establish a shared view of this?' or 'can we agree a way forward?'
- Questions that can be addressed to the whole team include, 'does anyone have experience of this issue?' or 'does anyone have any ideas on how we could deal with this issue?'
- In controlling the meeting avoid being overbearing or dominating the conversation, but firmly guide the participants in following the agenda and achieving the meeting's objectives.

- Interrupt politely and recommend a return to the agenda if conversation drifts too far from the agenda. Give a clear signal that 'we must move on'. Budget your time in the meeting and don't dwell on one agenda item to the exclusion of others.
- At the end of the meeting ask if there are any further items under any other business (AOB). If there are a large number of items there may not be time to consider them and they will have to wait until the next meeting unless urgent. Encourage members to inform you in advance of particular items that they want to discuss so that they can be placed on the agenda.
- End the meeting by summarising the main actions agreed. Ensure that everyone is clear about what they need to do and when, committing people to completion dates. Ensure you end the meeting on time.
- Agree times and dates for the next meeting.
- Remind staff how they can put items on the agenda for the next meeting.
- Thank everyone for attending.
- After the meeting complete the minutes and have them ready to distribute with the agenda for the next meeting. As a minimum include the following points in the minutes:
    - date and time of the next meeting
    - a list of those present and apologies
    - any corrections to previous minutes
    - matters arising from previous minutes
    - brief summary of discussion on agenda items
    - decisions reached and action required with staff who are responsible for action
    - timescales/deadlines for action.

Finally, consider what meetings cost in terms of time, travel and the salaries or wages of those attending. Ask yourself whether a meeting is essential? If meetings are poorly attended or participants consistently arrive late, or items constantly have to be deferred, what does this say about how the meetings are viewed and whether they achieve their purpose? Could other forms of communication be more efficient and serve the same purpose, for example using email or telephone calls?

Conclusions

Frontline managers are involved in a wide range of meetings and if these are badly organised and chaired this can lead to considerable time being wasted. Effective meetings can play an important role in communicating information and leading to collective decision-making. They can also have a valuable team development function. Being clear about the purpose of meetings and skilfully chairing them can improve their effectiveness and lead to a more satisfactory experience for the frontline manager and team members.

References

Coulshed V & Mullender A (2006) *Management in Social Work* (third edition). Basingstoke: Palgrave Macmillan.

Section 2

Chapter 10

Managing at the Front Line: Managing time and pressure

Key points

- Managing time, as a finite resource, means focusing on priority tasks.
- Managing time is about using techniques that enable time to be used more productively and concentrating on those tasks that are urgent and a priority.
- Time and pressure are intimately linked and there are a range of actions that can be taken to reduce pressure.
- How the frontline manager handles pressures and demands on them can help staff feel more confident in their manager.

Key words

Priorities; pressure; use of time; recognising stress; coping mechanisms.

Managing time

Managing in social care means working in a context where the pace and range of demands on managers and staff increases with each new initiative and the rate of change has similarly increased. Frontline managers find themselves having to respond to a continuous stream of activities that need their attention. The time available to meet these activities is a fixed resource. It is totally inelastic. Unlike other resources, if time is wasted it cannot be replaced. If managers are to create more time it has to be made through the efficient management of this finite resource.

Time management is essentially about priorities. It is recognising what needs to be done and making that a priority while at the same time saying 'no' to some activities, delaying others and delegating work to staff where appropriate. In addition, it is important that managers check that what they are doing reflects their priorities periodically, as these will change and develop over time. Although much of a manager's day-to-day work will be concerned with responding to the demands of the staff team and senior managers, it is important that frontline managers avoid reacting to others and don't lose sight of their own objectives and priorities.

Time management is about managing your time to achieve the things you want to do and that need doing. It's about having clear goals, planning the use of your time and how you approach the use of time available to you. To manage your time you need to know how you use it, then you can change what and when you do things so that you achieve more of your most important tasks.

Time and pressure are also intimately linked. If you are unable to manage your time it is likely that you will constantly feel under pressure and find that you are not able to complete your priority tasks. This will also be apparent to members of your staff team who will be less likely to seek your guidance or support if you always appear to be under pressure and panicking. Managing time is not only important for your own health but will also be communicated to your team.

Frontline managers who responded to the survey identified how they used their time and found that they were focused on seven main activities:

1. *'responding to staff who want support or information'*
2. *'arranging, chairing and attending meetings with their own staff and managers'*
3. *'reading and writing reports'*
4. *'responding to the demands of my manager'*
5. *'writing and responding to emails'*
6. *'receiving and making telephone calls'*
7. *'travelling to meetings and visiting staff working in different locations'.*

Having a disciplined approach

It is easy to be lured into spending too much time on things you enjoy doing and avoiding the things you do not like. It is important that you spend your time on the things that matter most, even if you do not enjoy some of the activities. This means you have to be focused, otherwise you will be wasting your time (Harrison, 2006).

More specifically it is useful to use your diary to plan ahead, making a list of activities that you need to complete each day and reviewing them at the end of the day. You should ideally plan your activities in a quiet period at the beginning of the day.

Other things you can do to improve your use of time include the following:

- Filtering emails that it is not necessary to respond to, and avoiding sending emails to others unless it essential they receive the information.

- Making meetings more efficient (see Chapter 9).
- Reducing your travelling time by more efficient use of new technologies to communicate with colleagues.
- Screening telephone calls and not allowing interruptions unless these are urgent when you are dealing with reports or correspondence.
- You can also increase your efficiency by creating a filing system that enables you to retrieve information quickly.
- Remember though that you and your staff need some time in the day to 'do nothing', whether it is socialising over drinks first thing or casual conversations about the work. These play an important part in staff relaxing and also building team identity (Hearn *et al,* 1992).

Identifying priority tasks

It is helpful to identify tasks in terms of their importance and urgency so that you can categorise them and decide what action to take. The following matrix diagram places tasks in whichever box is most appropriate.

1. Important and urgent	**2. Less important, but urgent**
3. Important, but not urgent	**4. Less important and not urgent**

What this diagram is describing is that there are activities that have levels of importance and urgency and you can place your tasks in whichever box is appropriate in relation to your priorities.

Box 1 is the tasks you do today.
Box 2 is the tasks you do today.
Box 3 is the tasks you plan time to do.
Box 4 is the tasks you do if you have time.

If the tasks in Box 4 keep hanging on your list time after time, ask yourself whether they really need to be done at all. If you consider they should be done then give them greater priority and complete them.

If you are focusing on those tasks in boxes 3 and 4 you may be neglecting activities in boxes 1 and 2 and your activities may need to climb up the order of priority. What this activity does is to help you focus on the important activities and avoid being diverted into other activities, no matter how pleasant, that in the end are unimportant.

Self-management

Self-management of your time is an important way of preventing stress-related illness, with the capacity to reflect on what you do on a day-to-day basis and to draw boundaries around the energy you have available for your career and that available for personal and family commitments. Some practical suggestions for managing your time more effectively from Coulshed and Mullender (2006) include:

- taking time before a meeting or telephone call to think what you want from it and keep this in mind throughout
- keep a note of everything you need to ask people so that you can cover it in one phone call or email
- arranging meetings in your office rather than travelling to someone else's
- shut your office door when you want to work undisturbed
- don't feel guilty about working at home
- allocate time for administrative tasks in your diary
- allow yourself time to 'walk the job'
- set aside time slots for specific activities such as record keeping, telephoning or emailing
- if you have to travel by public transport use the dead time for reading
- try to handle documents once, either deal with it, file it, or bin it
- control how others control your time.

Things to avoid

Avoid taking on more work than you can comfortably handle out of the mistaken belief that you are pleasing your manager, or meeting your need for approval, or responding to a culture that expects managers to consistently work late. This is a symptom of poor workload and bad time management and a lack of appreciation of the consequences can be damaging to the health of staff. Similarly avoid habitually over-cramming your diary with activities so that you have no time to chat or think. Time spent chatting to your team members is valuable interaction.

Analysing how you use your time

If you find that you do not control your time and you do not complete important tasks it is instructive to keep a time record for a week, jotting down each activity you are involved in for each hour of the day and analysing this at the end of the recording period, identifying what you spent most of your time on and how these tasks relate to your main priorities and whether you are spending your time on the most relevant activities.

This will help to reveal how much time is spent responding to the pressures of the job rather than dealing with important tasks related to your work objectives and developing new initiatives. Whether or not you decide to keep a time record it is important that you periodically analyse your time to make sure you are not spending time on activities that are unproductive.

Managing pressure

The pressures on managers grow as more and more tasks are devolved to the front line and the range of responsibilities increases. Frontline managers often describe their work in terms of constant pressure and not being able to complete any task satisfactorily. Although a certain amount of pressure can be stimulating, when it becomes too great it can easily turn into stress and can impair your effectiveness as a frontline manager.

Stress is an occupational hazard in social care, which is likely to affect everyone at some point. Social care has some features that make stress more likely (Harris & Kelly, 1992). These include:

- the relentless nature of human services
- the emotional demands of the work
- the undermining of self-worth by external criticism
- the exposed position of leadership and management in the service.

Stress occurs when the demands and the pressures of the work and non-work exceed the mechanisms we have for coping. Barlow (1990) identifies categories of potential stress, including:

The job itself

- too much or too little work to do
- having to work long and unsocial hours
- too much or too little variety in the work
- physical working conditions
- time pressures and deadlines
- factors outside your control
- changes in working practices and systems.

Your role in the organisation

- lack of clear objectives
- failing to delegate
- being constantly available
- having to make difficult decisions
- practitioner and manager demands/conflicts.

Your work relationships

- managing and supervising the work of staff
- lack of support by the team
- feeling isolated
- lack of positive feedback by your line manager
- personality clashes
- poor relationship with staff.

Organisational climate

- inadequate guidance and support from line manager
- lack of consultation and communication
- staff shortages/high turnover rates
- inadequate feedback on performance
- insufficient resources to work with
- poor morale or organisational climate
- demands of service users/carers/pressure groups
- pace and complexity of change.

Work/domestic interface

- taking work home
- not being able to 'switch off' at home
- spouse's/partner's attitude to your work
- demands the work makes on home life and the domestic situation
- absence of emotional support
- having few activities outside work
- putting your career at the expense of your home life.

Things you can do to reduce pressure and that will make you more effective

- Set yourself work objectives so that you have a plan and direction for your work.
- Take the initiative where possible so that you can have greater control of your work and don't let others control it.
- Set boundaries to your role and ensure that others know them so that you do not get overloaded with activities that are not your responsibility.
- If colleagues or your manager are overloading you, discuss with them how it is affecting you, and try to work out a better balance of activities.
- Look after yourself, developing some stress reduction techniques and having proper drinks and lunch breaks.
- Try to reduce tension and lowering damaging arousal through relaxation, play, exercise or hobbies.
- Use positive thinking recognising the destructive self-critical evaluation and destructive self-references.
- Be aware of the value of self-disclosure: sharing your feelings informally with family, friends or colleagues, or formally with professional counsellors.
- Learn to be assertive in your responses and don't feel that you have to meet everyone else's needs and neglect your own. Use direct, straightforward expressions of your opinions, beliefs, values or ideas. Avoid aggressive or non-assertive responses.
- Use support from your colleagues and develop a group of people who are willing to help you through stressful situations.
- Monitor your stress by being aware when tension builds up, and the events or situations that trigger stressful responses.

- Look after yourself by monitoring your health and keeping fit. Avoiding dependency on alcohol or smoking when stressed.
- Be kind to yourself, don't beat yourself up if things don't always work out.

Managing under pressure

Where there is intense pressure on a service, with the primary aim being to get through the work, rather than concentrating on effectiveness or creativity, this can provide feelings of exhilaration and the flow of adrenaline, but only if they are occasional rather than routine. Where this becomes the norm it can lead to a loss of effectiveness both for the service and its staff. Your job as frontline manager is to prevent this happening on a regular basis. One way is to create a sense of calm and control, for instance, making clear judgements under pressure and not panicking, which is communicated to the rest of your team.

Creating an aura of calm

An important aspect of management is presenting an impression of calm, particularly in pressurised situations, or when decisions have to be made. Although you may feel anxious inside and are wondering how to cope with your workload, if you can suppress the signals that reveal your anxiety and that might disconcert staff then this can create a good impression and, in turn, inspire confidence in your management (Martin & Henderson, 2001). Martin and Henderson suggest that one way in which you can create the impression of being in control is to set aside time to meet with staff and discuss their problems in a relaxed way. Again this suggests that you are able to manage your time effectively in order to create space in your working day to meet staff. If you can demonstrate that you manage your time in a way that helps you to be effective and in turn support others, it will help them respect and value their time as well as yours.

Where you are working under pressure you set the tone by clearly stating the priorities, discussing them with staff and explaining this situation to your manager and other colleagues who may be affected. This may also need to be explained to service users. This is an important action for you to take as others can be more reasonable about delays if they understand the reasons for it (Hearn *et al,* 1992).

Conclusions

The demands of social care services mean that frontline managers will inevitably experience pressure on their time and, if not managed, this can lead to the damaging effects of stress. The recognition of time as a resource and its management is essential to understanding your priorities and developing strategies to use this resource as effectively as possible. How you manage pressure will also have an impact on staff and their confidence in your management. Discussing the sort of issues that cause pressure with your manager and staff and getting their support can be helpful as a coping mechanism.

References

Barlow A (Ed) (1990) *Transitions: A self-assessment and development package for managers in social services*. Luton: Local Government Management Board.

Coulshed V & Mullender A (2006) *Management in Social Work* (third edition). Basingstoke: Palgrave Macmillan.

Harrison P (2006) *Managing Social Care*. Lyme Regis: Russell House Publishing Ltd.

Harris J & Kelly D (1992) *Management Skills in Social Care. A handbook for social care managers*. Aldershot: Ashgate.

Hearn B, Darvill G & Morris B (Eds) (1992) *On Becoming a Manager in Social Work*. Harlow: Longman.

Martin V & Henderson E (2001) *Managing in Health and Social Care*. London: Routledge.

Section 3

Managing the service

Section 3

Chapter 11

Managing at the Front Line: Managing quality and standards

Key points

- Quality is central to the delivery of services that consistently meet service users' needs and comply with regulatory standards.
- The government's modernisation agenda has introduced a wide range of quality initiatives and processes.
- Understanding quality concepts and techniques can help the frontline manager to select those processes that can be used to improve a service.
- Involving service users, carers and staff can be an effective way of improving services, particularly where changes are necessary.

Key words

Quality; modernisation; minimum standards; quality concepts; consultation; service improvement and change.

Quality and standards are of primary importance to users and carers. Quality is also important to a service, its image, its reputation and its long-term success.
The development of quality and the maintenance of standards are crucial to improving a service and reducing the costs of errors and mistakes and means that time does not have to be spent correcting errors and avoiding the costly expense of potential legal action against the service. It is also important to staff motivation and commitment, as staff will take pride in working in a service that is acknowledged as high quality and where their contribution is recognised and valued.

Quality has been described as the provision of services that are fit for purpose and at a reasonable cost, taking into account the choices available to the people who use them (Coulshed & Mullender, 2006). In *A Quality Strategy for Social Care* (DH, 2000) quality is concerned with changing and improving services so that they are more accessible and consistent and are delivered by a workforce that has the skills to meet the needs of people who use them. Ultimately, quality is about how well services meet the needs of those who use them and, in turn, how those who manage services can continually improve them to achieve this goal.

The emergence of the concept of quality in public services

Quality has a long history with its origins back in the early 20th century and the improvement of industrial processes. For social care services its history is much shorter and has its roots in the late 1980s and early 1990s when the Conservative government introduced reforms that would transform social work and social care services. These reforms, in both the NHS and local government, saw the adoption of markets to improve the effectiveness and quality of services through opening the public sector to competition. The adoption of management ideas, concepts and practices from the private sector included techniques such as quality control, quality assurance and total quality management, which had up to that time been the preserve of business and industry.

This was also the period when services were restructured to separate the commissioning and purchase of services from its provision, which began the transfer of provision to the independent sector. Along with the separation of the purchasing and providing functions, came focus on the quality and costs of services purchased from the independent sector, with the growth in the regulation and control of these service providers. This period saw the introduction of inspection units in local authorities and the gradual development of quality assurance systems in social services departments and the independent sector.

Labour government and modernising social care

With the election of the Labour government in 1997, a further stage of reform of social care services was initiated, focusing on the 'modernisation' of public services with the introduction of a wider range of processes with quality of services and outcomes for users and carers central to these changes. Rather than rejecting the market-based models of its predecessors, the Labour government strengthened and extended the role of markets in social care, alongside a range of new policy initiatives that focused on raising the standards of social care services with a wide range of new regulatory and inspection methods introduced to control and assess the performance of services. Some of the main changes that have relevance for social care include the following.

- **Best Value** was a government initiative introduced in 2000, which replaced compulsory competitive tendering (CCT) introduced by the previous Conservative government. It required local authorities to systematically and continuously review and evaluate their services with the intention of promoting better standards and demonstrating quality and value for money. The Audit Commission checked that local authorities challenged themselves sufficiently when carrying out the Best Value reviews. Local authorities were also required to publish an annual Best Value Performance Plan setting out how they performed over the past year with targets and future plans. Best Value has recently been replaced by a new National Indicator Set from April 2008. See www.communities.gov.uk.

- **National Care Standards Act (2000)** established the National Minimum Standards (NMS) for social care services and created the Commission for Social Care Inspection (CSCI), which regulated residential and nursing homes, domiciliary services and adult placements, along with children's residential homes and special education, fostering and adoption agencies. (CSCI has now been merged with the Healthcare Commission and the Mental Health Act Commission as the Care Quality Commission introduced in April 2009). See www.cqc.org.uk.
- **Star ratings for social services** (2001) brought together evidence from the Audit Commission's judgement of performance based on a scale of zero to three stars. This is part of the government's comprehensive Performance Assessment Framework (PAF), with the ratings of local authorities intended to improve public information about the current performance of services and their capacity for improvement. Authorities that perform well have more freedom in the way they use their centrally provided resources with reduced levels of inspection and monitoring. Authorities with zero stars are more frequently monitored and have support arrangements put in place to enable them to improve their performance. See PAF for social services on ratings on www.cqc.org.uk.
- **Fair Access to Services** (2002) was introduced to address the issues of the 'postcode lottery' of service provision and established four categories of eligibility for services ranging from critical to low, with the needs of individuals and the risks to independence assessed against this criteria. See www.dh.gov.uk.
- **General Social Care Council** is the governing body for social workers and social care staff and has established the national codes of practice for all social care staff and their employers with the introduction in 2003 of the Social Care Register. The GSCC is also responsible for regulating and supporting social work education and training including accrediting social work qualifications at both qualifying and post-qualifying levels and also quality assure all social work courses. They are also responsible for the development of National Occupational Standards for staff working in social care as seen in the occupational standards for social workers and the National Vocational Qualifications such as the Registered Manager Award. See www.gscc.org.uk.
- **Social Care Institute of Excellence** collects and synthesises up-to-date knowledge and information on what works best in social care services, and disseminates this knowledge base to improve practice. See www.scie.org.uk.

Importance of standards

Standards are important as they act as a minimum guarantee that users should be able to expect from a service. As a frontline manager your responsibility in relation to standards is to publicise the standards for your service to ensure that all staff fully understand these and what they are required to do to meet them. This will mean providing training and support to ensure that all staff can practice to the required standards, checking that standards are always met and acting quickly to correct practices where standards are not being met. For frontline staff, who are the critical actors in the service chain and who have to deliver quality standards, they must be able to rely on leadership and management with systems in place to ensure that they feel supported in their practice. Your approach to monitoring standards needs to ensure that staff do not feel they are being 'caught out' and that there is a blame-free culture where the quality system is a participative process committed to ensuring that staff feel that they can discuss their practice in an environment where the focus is on identifying ways of improving practices rather than identifying errors.

Concepts of quality

Quality has a whole range of concepts with their own jargon that can be mystifying to those who first confront it. The following concepts are important to understand as they relate directly to what happens in social care services.

- **Quality control** originally referred to the inspection of industrial products to ensure that they met a predetermined standard. Any that failed to meet the standard were rejected and consequently carried considerable costs. To translate this concept to social care services, it is seen in the CQC inspections where the inspectors assess the performance of a service against a criteria based on the National Minimum Standards for that service. Quality control tends to focus on those things that can be easily observed or counted as it is measuring success against a pre-determined criteria. It is less successful in more subtle areas of personal perception and experience of services, hence the complaint of many managers that it ignores the atmosphere and the quality of relationships between staff and service users.
- **Quality assurance** focuses on each stage of the production or service process to eliminate faults and involves designing and planning services in a way that ensures that the service will meet the agreed standards without errors. It is intended to provide the correct service first time and avoid the waste and inappropriate behaviour that occurs when there are errors (Patel, 1994). It means considerable work for managers and staff in developing systems that ensure all parts of a service are robust and deliver what it says they will do.
- **Total quality management** is an organisation-wide process with the intention of creating a total organisational approach to quality. It focuses on how all aspects of a process contribute to the quality of a service. All staff are trained in quality techniques and quality becomes the responsibility of all staff. It stresses the importance of teamwork, systems, communications and continual improvement to ensure that quality standards are consistently achieved. It attempts to avoid the fragmented approach to quality found in many organisations and the continual search for new quality concepts rather than concentrating on current approaches (Dale, 1999).
- **Continuous improvement** is the long-term work in continually improving all aspects of a service, driven by a vision of what a service is trying to achieve, with managers having a detailed knowledge of their service with constant problem prevention and improvement activities undertaken (Dale, 1999).
- **Costs of quality** include the costs of introducing and maintaining continuous quality improvement and the management and staff time needed to ensure a service meets

its quality standards. The second area of costs are those incurred as result of making mistakes and having to correct work later. These costs can include failures in the delivery of services, care process and products used. Although it can be difficult to attribute costs to some failures of quality it is important that managers identify problem areas so that corrective action and improvement can be undertaken.

Limitations of quality systems

It is important to recognise that the operation of some social care services is less predictable and more turbulent than the organisations that many of these quality systems were originally designed for. Likewise, when these systems are introduced into public services they are often designed by consultants or senior managers at the behest of politicians and can conflict with managerial and professional interests and values and not achieve their objectives in improving services (Hafford-Letchfield, 2007).

The frontline manager's role in quality

As a frontline manager you have a series of opportunities to improve the quality of your service, primarily through your leadership of the service, where you act as a role model for the culture of quality so that your ideas permeate the organisation (Hafford-Letchfield, 2007). Hafford-Letchfield suggests three activities that can increase your influence – *listening to* people and demonstrating caring, *teaching* where you transmit positive values and *facilitating* where you give on-the-spot help.

Discussion with frontline managers from group care services illustrates the demands that their work to improve standards are made on them.

> *'The main challenge I faced was putting new systems in place to improve standards as I felt this had been neglected and bad practices had crept in.'*
>
> *'I had to get directly involved in improving standards by changing routines for the benefit of users. I had to be honest and open with staff and accepted that the changes wouldn't suit everyone.'*
>
> *'It's hard work changing people's attitudes when they have been used to working in their own way for years, but you've got no option – you know you've got to change things.'*

Leading and managing your service means ensuring that it meets the standards set. First, this requires understanding and implementing the National Minimum Standards for your service and working with the CSCI to ensure that your service meets the inspection requirements. You also need to support members of staff to ensure that the service meets the GSCC requirements by ensuring that the Codes of Practice, which describe standards of professional conduct and practice are promoted in the service (GSCC, 2005). You should also ensure that staff are able to access social care education and training opportunities so that their skills match the demands made on them.

Consulting with service users

We have seen that improving services for users is central to quality and it is here that you can influence the quality of your service, through consulting with service users about what they want from services, acting on their demands and recommendations where possible and letting them know when you are not able to meet their needs and the reasons why.

You need to consult regularly with users to ensure that their expectations are being met and identify changes and needs that may require you to revise standards. Demanding service users with high expectations can be very effective change agents. If we look at quality from the service user's point of view, their perceptions, expectations and experience of a service needs to be taken into account. Some useful questions suggested by Lawrie (1992) to ask are:

- Is there a gap between what users expect of the service and what the service actually delivers?
- Is there a gap between service users' expectations and what service providers think users' needs and wants are?
- Is there a gap between how the service is managed and how it actually performs?

What you can learn from asking these questions is demanding and needs considerable time and effort, but it is essential and worthwhile as it focuses on the needs of service users. It can also cut the costs of a service by eliminating unnecessary functions and processes that do not contribute to meeting users' needs. As a considerable amount of management time can be spent reacting to things that repeatedly go wrong, quality assurance has considerable benefits. Having robust quality systems in place can improve the standing of a service and its credibility in the eyes of purchasers and potential users.

Using research evidence to support change

The growth of the internet has provided you with the opportunity to access a wide range of social care websites that provide information on good practice in social care, findings of research (for example, www.scie.org.uk) and reports of inspections that can suggest ways that you can improve your service. With the increasing importance of evidence-based practice it is essential that you exploit the use of information technology to support improvements to your service. Similarly, encourage your staff to use these sites so that they become familiar with the range of information available and use these resources to develop their knowledge and skills about practice. Access to information can begin to help staff feel more comfortable with change, which is essential if services are to improve.

Promoting continuous improvement

Your task is to develop and maintain the quality of your service and this is most usefully achieved through the use of the quality improvement cycle (Thomas *et al*, 2003). There are three stages to this cycle.

1. Being clear about the standards for your service and what you need to improve.

2. Working with your staff to ensure that they understand the need for improvement.

3. Collecting ideas for change, both drawing on your own experience, those of your staff and suggestions from service users and carers.

Persuading staff to change their practice is not always easy and using an evidence-based approach can be a powerful way of convincing staff of the need for change – your argument could include:

- who or what is driving the change
- why the change is necessary
- how it will benefit service users
- how it will benefit the service.

Quality improvements are not achieved overnight, which was the weakness of many of the original quality improvement programmes introduced in the 1980s that assumed top-down instructions and dissemination of quality techniques would be quickly adopted by staff (Joss & Kogan, 1995). They failed to understand the need to engage with staff and communicate the importance of quality and how staff themselves could develop initiatives that would be more likely to be successfully implemented. Quality improvement requires more than quality systems and toolkits being promoted by senior managers and consultants.

Managing the long-term process

Quality improvement is a long-term organisational development process. It involves leadership, communication, discussion of ways of improving services and the building of a culture that values excellence. It means engaging with service users and staff to provide ownership of the quality agenda and where every member of staff feels committed to ensuring that they provide the highest quality service possible. To achieve this means developing a culture that values the contribution of everyone and where managers listen intently to their ideas and suggestions for improvement. Avoid making your approach to quality too complex. Focus on those aspects of the service where it is important. Do not let it degenerate into a time and paper consuming exercise that pulls staff away from delivering the service at the front line and thus reducing performance (Lawrie, 1992). In the end, quality is not about jargon and rhetoric, it is about the long-term persistent focus on improvement that means day-by-day attention to the minutiae of social care to ensure that a service continually meets the required standards.

Conclusions

The importance of continually improving the quality of a service is a central part of the frontline manager's role. Understanding the importance of service users' views about quality allied to the requirements of national standards provides the basis for the use of a range of quality techniques and processes. Improving service on an ongoing basis means continuously modifying practice where necessary to ensure that a service meets the needs of service users.

References

Coulshed V & Mullender A (2006) *Management in Social Work* (third edition). Basingstoke: Macmillan.

Dale BG (Ed) (1999) *Managing Quality* (third edition). Oxford: Blackwell.

Department of Health (2000) *A Quality Strategy for Social Care*. London: Department of Health.

General Social Care Council (2005) *Codes of Practice for Employers of Social Care Workers*. London: GSCC. Available from: www.gscc.org.uk.

Hafford-Letchfield T (2007) *Practising Quality Assurance in Social Care*. Exeter: Learning Matters.

Joss R & Kogan M (1995) *Advancing Quality, Total Quality Management in the National Health Service*. Buckingham: Open University Press.

Lawrie A (1992) *Quality of Service, Measuring Performance for Voluntary Organisations*. London: NCVO/Directory of Social Change.

Patel A (1994) Quality assurance (ISO 9000) in social services departments. *International Journal of Public Sector Management* **7** (2) 4–15.

Thomas A, Mason L & Ford S (2003) *Care Management in Practice. For the registered manager award*. Oxford: Heinemann Educational Publishers.

Section 3

Chapter 12

Managing at the Front Line: Managing poor performance

Key points

- Managing poor performance can be one of the most difficult areas of practice for the frontline manager.
- It is important to be clear about what constitutes poor performance and some of the reasons why staff fail to perform to the required standard.
- Frontline managers need to understand the difference between capability and misconduct and how to address them.
- Action to resolve performance issues can range from informal discussion through to formal disciplinary procedures.
- It is important to seek advice and support from a senior manager and/or HR staff when dealing with serious performance issues.

Key words

Poor performance; reasons for poor performance; capability; misconduct; taking action; informal and formal procedures; DESC script; disciplinary procedures.

One of the most difficult areas facing frontline managers and an area of practice that generates high levels of anxiety is how to manage persistent poor performance, and the frustration when intervention fails to improve performance. For newly appointed managers learning the skills of managing staff who are clearly not performing to the required standard can be one of the most challenging management problems they will face. This is also an area of practice where managers often felt on their own with little back-up from their own line manager or wider the organisation.

> *'One of the most difficult parts of my job has been dealing with a member of staff whose poor performance is damaging work with users and morale in the team. In spite of numerous requests for help I've had little support. I'm expected to manage the situation even though it's been going on for years. I seem to get no help from my manager.'*
>
> Manager, local authority residential service

Not dealing with poor performance on your own

Support from your line manager and the human resource department staff (or consultation with an external HR adviser and the use of websites such as ACAS) is essential if frontline managers are not to be left unsupported to deal with poor performance on their own. You can prepare yourself for dealing with poor performance by familiarising yourself with your organisation's procedures on poor performance, and by doing some research using websites and ensuring that you record and reflect on instances of poor performance, so that you present a clear and concise case when you need the advice of others.

Valuing of meeting with other managers

Another strategy you can employ to avoid feeling isolated when dealing with poor performance is to meet collectively with other line managers to discuss performance problems, share experiences, ideas and approaches and encourage and support each other. When dealing with complex issues of performance it can be particularly difficult to think about solutions on your own and in isolation, so consulting with others is a vital source of help.

Managing performance has always been an integral part of the management role but has increased significantly with the increasing influence of service users over the quality of service they receive. The introduction of performance management with its intention of relating individual and team performance to wider service and organisation targets, the introduction of National Minimum Standards for services and the regulatory framework forming the basis of CQC inspections of services, and procedures that are designed to ensure that service users are safeguarded from abuse through procedures developed by the Association of Directors of Social Services (ADSS, 2005).

What is poor performance?

Poor performance is the gap between required and actual performance, with staff having a contractual responsibility to achieve a satisfactory level of performance. Employers must set realistic and achievable standards and ensure that when staff are appointed they understand what is required of them, through job descriptions, induction and discussion about their duties with their manager. Examples of poor performance can include, poor standards of practice, an inability to cope with a reasonable volume of work, lack of skills and consistently not achieving agreed and realistic targets. Once the reason for the gap is identified it provides you with an opportunity to address the problem.

Where there is a shortfall in performance this needs to be pointed out to the member of staff and reasons for the shortfall identified. There are a wide range of potential reasons for poor performance, and as the frontline manager you are ideally placed to identify with the member of staff, whether there are organisational factors that are the causes of their poor performance, or whether the problem lies with the member of staff who is unable to meet the required standards. Some of the causes of poor performance include:

- an unsuitable appointment, not uncommon where there is pressure to fill posts
- a job description that is too wide or difficult to achieve
- lack of clarity about the range of responsibilities
- there has been no induction or the induction was poor, with training needs not identified
- there has been poor use of the probationary period, a lack of review and this period was not extended where there were doubts about the person's performance
- irregular or poor-quality supervision
- no annual appraisal or unclear targets
- lack of resources to do the job
- lack of skills to do the job
- poor workplace relationships with manager or peers
- a mismatch between the culture of the organisation and the member of staff
- the member of staff is experiencing personal problems that are affecting their performance
- the member of staff is poorly motivated.

Wider implications of poor performance

Poor performance on the part of a member of staff has a number of consequences for social care teams. Awareness that a colleague is performing poorly can have a knock-on effect on the wider team, lowering morale and placing additional demands on others to ensure that service standards are met. Where poor performance is not addressed this can send a message to other staff that achieving standards are optional and that poor-quality work is acceptable. Other consequences can occur where poor performers cover up their own shortcomings by bullying or manipulating other staff. Clearly, managers have to take action to address poor performance before its damaging effects impact on service users and reduce the morale and performance of the service.

The comments from a manager of group care services sums up the impact of poor performance on her team:

> *'It can have a domino effect on fellow care staff and their motivation. The service users suffer and it causes extra stress on managers. If it's not dealt with other staff feel they can get away with it. Staff loose faith in the manager if they don't act.'*

Questions to ask yourself

The Social Care Institute for Excellence (SCIE, 2003) provides a guide on managing poor practice for frontline managers, which is helpful in clarifying the questions that managers need to ask themselves when concerned about poor performance. Are you clear about your own, your organisation's and any professional expectations about the following.

- What is acceptable behaviour at work?
- What are the performance standards for your service?
- How will you negotiate differences of opinion on professional standards?

- Are you confident that these expectations are explicit and have been shared with the team you manage?
- Are service users and carers clear about the standards they can expect from your staff?
- Are you confident that you will be told by team members if they are concerned about the behaviour of a colleague and standards of practice?
- Does the whole team have a stated and shared responsibility for ensuring appropriate standards of behaviour and practice?
- Do they collectively safeguard the interests of service users and carers?

In managing performance there are a number of ways in which you can ensure that staff are meeting the service standards, including the use of the National Minimum Standards and CQC standards judging their performance through direct observation and/or completion of paperwork and setting targets that will meet or exceed service standards.

Capability

Many organisations have now developed capability procedures to deal with poor performance and these are not seen as disciplinary procedures as the emphasis is on helping the member of staff to perform more effectively in the future. This is done through supervision, training and support and the member of staff is given a reasonable time to achieve the required standard. Capability is where the person does not understand what they have to do or is not capable of doing it. The inability to achieve the standard required may be the result of a lack of knowledge, skills or ability, rather than misconduct. For example, is the worker failing to fully complete the service users' care plans and has this happened on a number of previous occasions? Consideration should be given as to whether the poor performance is due to inadequate induction, training, supervision or some other factor. When first employing a new member of staff, does your organisation ensure that it clearly explains the standard of work required so that the member of staff is left in no doubt about what is expected of them? Similarly, does the job description accurately convey the main purpose of the job and the tasks involved?

In the formal part of a capability procedure, which will vary from organisation to organisation, it would usually involve a series of meetings with the employee giving them clear targets to meet, with failure to meet those targets resulting in warnings and potential dismissal. Again, using capability procedures should only be undertaken with the support of your manager and the HR department or an adviser.

If there is no improvement in performance, in spite of attempts to support the member of staff through training activities and increased support and supervision, it may be necessary to move to the stage of the formal capability procedure, which could ultimately lead to dismissal if the person is not able to improve their performance. This is a serious step and these should only be taken in conjunction with the support of your manager and your HR department.

Misconduct

Misconduct is where the person is knowingly not doing what is required of them. With misconduct the quality of the member of staff's work may not be in question, but their behaviour is unacceptable in relation to the service's standards of conduct. For example, is the member of staff persistently late for their shift, even though they have been reminded informally about punctuality on a number of occasions?

Organisations may have written guidance that differentiates between minor and gross misconduct. An example from a service for people with learning difficulties differentiates minor from gross misconduct in the following ways.

Minor misconduct

- Poor or unsatisfactory performance
- Poor timekeeping
- Failure to follow absence procedure
- Rude or uncooperative behaviour
- Misuse of property or equipment
- Poor appearance
- Discourtesy
- Failure to comply with rules relating to public statements

Gross misconduct

- Theft or other fraudulent behaviour
- Sleeping on duty
- Unauthorised absence from work
- Persistent lateness
- Making misleading statements
- Incapable of performing duties due to alcohol or drugs
- Sexual advances/relations with a service user
- Gross indecency
- Use of abusive language or behaviour
- Acts of harassment
- Abuse of the sick pay scheme
- Refusal to comply with reasonable management instructions

Instances of minor misconduct are usually best dealt with in an informal manner, pointing out to the member of staff what they are required to do to improve their conduct. Several managers described how an act of misconduct by a member of staff was often inadvertent and after a quiet and confidential word there was no reoccurrence of the misconduct. Where there are further acts of misconduct then more formal action should be taken.

Where you are concerned about poor performance you have a number of options open to you. The first step is to raise your concerns with the member of staff, for example in a supervision session where it is legitimate to raise your concerns without it being seen as part of the formal capability or disciplinary procedure.

Absence management

Unauthorised absence from work is a serious cost to a service, both in direct costs (salaries and wages) and indirect costs (risks to service users, impact on a service, additional pressures on colleagues and management time taken in dealing with absences). Research into unauthorised absences show a pattern of regular Monday and Friday absences, sickness on return from holiday and absences related to local events. Absences can also be related to unpopular shift times or colleagues (Evans & Walters, 2002).

Frontline managers have an important role in reducing absence levels using prompt and fair action. Organisations that train their frontline managers in absence management show a reduced level of unauthorised absences. There are a number of actions that organisations can take to reduce absences, including having absence policies and procedures that are communicated to staff and training frontline managers in monitoring absences. Procedures include staff being required to inform their line manager directly by telephone when they are reporting sick and the reason for their absence, and managers meeting with staff on their return to work to discuss their absence, with intervention and support for staff who have problems with attendance. The actions of the frontline manager need to be beyond reproach with the service demonstrating that it acts fairly and responsibly, and respects the member of staff's rights under employment legislation.

> *'Unauthorised and regular sickness absence was very costly for our service with agency cover needed and the added stress for other members of the team. Since we introduced a policy on absence, with staff required to ring their line manager on the day of the absence, and the line managers being responsible for arranging a return to work interview with staff who have been absent, it has significantly reduced the amount of days lost.'*
>
> Frontline manager, mental health service

Capability and conduct: taking action early

You are now in a position to distinguish between capability and conduct. It is important to understand the source of the poor performance and whether it is capability or conduct, although there may be an overlap of the two.

It is important to identify and address poor performance at an early stage before it causes major problems for the quality of service and has an effect on other team members. Managers spoke of the importance of promoting service standards and regular supervision as two ways of being much clearer about performance standards and making it easier to know when to discuss this.

> *'Poor performance is something I try to address in supervision. It needs to be addressed early before things become a real problem. It's been difficult dealing with it (poor performance) when it's been ignored or swept under the carpet by a*

previous manager. It's crucial to keep detailed records of the problem and what you have said to the member of staff as you will need them to back you up later if you can't resolve it at an earlier stage.'

Team leader, adult services

If you have a member of staff whose performance concerns you, what action do you take?

- You can raise the issue of the staff member's performance in a supervision session or an appraisal meeting with the aim of resolving the issue of poor performance at this stage.
- Ensure you keep detailed records of your discussions with the member of staff as these records may be important at a later date if you have to move to a disciplinary stage.
- Discuss your concerns with your line manager and keeping them up to date with progress in resolving the problem at an informal stage.
- Consult your HR department or alternatively familiarise yourself with the guidance on the management of poor performance contained in the ACAS handbooks.

Good practice in tackling poor performance

Good practice is important when dealing with poor performance. Ford and Hargreaves (1991) suggest the following steps in tackling poor performance:

- plan ahead and think through what you will say and how you will say it before meeting the member of staff
- share your concerns with the member of staff at an early stage – don't let the situation develop to the stage where it is impossible to take corrective action
- focus on the facts
- act assertively, not aggressively or non-assertively
- explain why you are raising the issue
- ask for a response – ask the member of staff for their view of what has happened and why
- identify the causes of the poor performance with the member of staff
- plan together to correct the poor performance, setting short-term objectives
- agree a date to review progress
- if the poor performance continues, begin to look longer term and explore likely consequences.

Preparing for the meeting with a member of staff: using the DESC script

Many managers commented on their anxiety on how they would actually communicate the problem to a member of staff. The use of the DESC script technique can be useful as a way of helping you think through what to say, to be clear and forceful, and to describe the problem in a non-judgemental way.

- Describe (D) what the member of staff is doing that is creating the problem.
- Express (E) why the behaviour is a problem – for you as the manager and for the organisation.
- Specify (S) what the member of staff should be doing instead.
- Consequences (C) of either succeeding or failing to change the problem behaviour.

The presentation should be thought through before meeting with the member of staff. It can also help to write the script out and practise it beforehand. This provides a mental checklist when you actually confront the member of staff.

What a member of staff hears

A useful way of checking out your presentation is to ask yourself the following questions.

- If I was the person receiving this script, would I be clear about what I am doing that is a problem, why there is a problem and what my manager wants me to do about it?
- Are the statements by the manager likely to put me on the defensive?

Some guiding principles

- Maintain your dignity and self-respect
- Seek to understand where the person is coming from
- Remember you can't change the person, aim to change the behaviour
- Express your own point of view and stick to it, do not be side tracked by other issues
- Avoid points scoring and work towards a solution

Gross misconduct

Your organisation should have procedures for capability and misconduct as there are statutory dismissal and grievance procedures. HR departments or trade union representatives will know the procedures well and you should consult them to ensure that you follow the procedures correctly. The consequences for not following the capability and disciplinary procedures correctly can be costly and embarrassing to an organisation and you should familiarise yourself with the procedures before taking any action. Many services, particularly smaller ones without HR departments consult the ACAS *Code of Practice* (2004) and use this as a basis for their procedures. In cases of gross misconduct there should be well-defined procedures, with the potential for summary dismissal. The advised route is to suspend a member of staff, with full pay, to establish the full facts by investigation. As a result of the investigation the organisation may then choose to take disciplinary action.

Disciplinary procedures

Where poor performance persists in spite of action to correct it, it may be necessary to use disciplinary procedures. These procedures should be seen as a means of correcting and improving standards of performance, not of administering punishment. As a frontline manager you will be expected to initiate these procedures when necessary and take part in disciplinary hearings, usually with the involvement of HR specialists or senior managers. Your role may include investigating an incident and you should look for guidance from your organisation with regards to how to do this in an effective but fair way. Larger organisations usually employ their own specialist HR staff who are familiar with disciplinary procedures, whereas small independent sector organisations tend to rely on external advice either from employment legal specialists or through the ACAS advice service. All organisations should have dismissal and disciplinary procedures as set out in the Employment Act (2002) and fully described in the ACAS guidance.

Conclusions

Dealing with members of staff whose performance is poor can be a major area of anxiety for many frontline managers. It is vital to establish the causes of poor performance before taking action. Taking action at the earliest stage possible once the cause has been identified is good practice. Frontline managers should not attempt to deal with serious poor performance on their own. It is important to draw on the expertise of the HR staff in your organisation and senior managers when misconduct is likely to lead to disciplinary action.

Useful resources

There are a number of useful resources on managing poor performance on websites. The following sites are all worth consulting for guidance on different aspects of performance issues and the legal requirements in relation to capability and disciplinary processes:

www.acas.org.uk
www.peoplemanagement.co.uk
www.personneltoday.com
www.scie.org.uk

References

ACAS (2004) *Code of Practice – Disciplinary and grievance procedures*. London: ACAS. Available from: www.acas.org.uk.

Association of Directors of Social Services (2005) *Safeguarding Adults: A national framework of standards of good practice and outcomes in adult protection work.* London: ADSS.

Evans A & Walters M (2002) *From Absence to Attendance* (second edition). London: CIPD.

Ford K & Hargreaves A (1991) *First Line Management: Staff.* Brighton: Pavilion.

SCIE (2003) *Practice guide 1. Managing poor performance*. London: SCIE. Available from: www.scie.org.uk.

Section 3

Chapter 13

Managing at the Front Line: Managing workplace conflicts and disagreements

Key points

- **Conflict and disagreement are common in all organisations and their consequences are dependent on their seriousness and how they are perceived.**
- **There are a wide range of workplace factors that can lead to minor disagreements and more serious conflicts.**
- **Different approaches to handling conflict can be used, although it is important that they do not leave the manager and staff compromised in their resolution.**
- **How conflicts and disagreements are managed and the techniques using in resolving them should be understood, and the frontline manager needs to develop skills in conflict resolution.**

Key words

Managing conflicts; bumps, bruises and battles; five approaches model; resolving grievances; prevention.

Alongside managing poor performance, dealing with conflicts and disagreements consumed a significant element of the frontline managers' time, based on those surveyed for this handbook. There are differing views on conflict and disagreement in the workplace and to a great extent it is the degree of the conflict and disagreement and how they are seen and managed that is important. Conflict is a reality in all organisations and for that reason should be managed rather than suppressed. Not all conflicts and disagreements should be seen negatively as their presence can lead to improved problem-solving and reveal issues around different values that can improve services.

Conflicts and disagreements

It is useful to differentiate conflicts from disagreements. Conflicts tend to reflect the structural features of an organisation where individuals or groups of staff are concerned about aspects of the service such as roles, professional relationships, service delivery issues, concern about working conditions or requirements to change practices. Whereas disagreements are largely interpersonal between members of staff and are often of a personal nature that can be unconnected with the work, although the close proximity of working relationships can exacerbate disagreements.

It is important to recognise that there will inevitably be individual differences among staff groups, particularly where colleagues are working in close proximity to each other and service users, for example in group care settings. You should also consider that your own personality and management style will be a factor in any interpersonal conflict.

Causes of conflicts and disagreements

Conflicts and disagreements have their origins in potentially wide-ranging sources:

- the way work is organised
- relationship between the individual and the organisation
- management style
- individual values and beliefs – which can be ideologically, career or job related.

Some typical conflicts and disagreements described by frontline managers

- *'Between the frontline manager and an individual member of staff related to performance issues or new working practices.'*
- *'Between members of staff where there is conflict about roles or role boundaries.'*
- *'Between two members of staff where there is personal animosity or other hidden issues.'*
- *'Between groups of staff, for example night and day staff in a residential setting over respective responsibilities.'*

These are the typical examples of conflicts that are part of everyday management and should, where possible, be dealt with informally rather than needing to move to more formal stages, such as disciplinary procedures, or where staff feel their concerns are ignored and not dealt with effectively, and use the grievance procedure to resolve a conflict.

Categorising conflicts

When thinking about conflict it is useful to categorise them in terms of their significance as this will help you decide how and when you intervene. Here, I have adapted some ideas from Goodwin and Griffith (2007) for resolving workplace conflicts in the following diagram.

Bumps These are disagreements that do not damage individuals' relationships or the service, and where staff work out their own solutions, at least for the moment. They can be a healthy sign that individuals with distinct personalities and viewpoints are trying to reconcile their differences.	You need to monitor this situation and ensure that it does not evolve into conflict. If the parties are mature and the disagreement is manageable there is no need to intervene. This can wait.
Bruises This is an emerging conflict that may have developed out of a disagreement and got out of control and beyond the ability of the parties to handle. This may involve a breakdown in communication and accusations being made about each other.	This is a conflict that has not fully erupted and you have the opportunity to step in and either eliminate it, or at least keep it from becoming a fully fledged battle. This needs preventative action.
Battles This is where true conflict has erupted between parties and its impact is having an affect in terms of lowered morale and performance, and creating disunity and discontent among staff.	This is where you need to use your understanding of the issues and the parties involved and your knowledge to resolve the conflict. This requires action. You you need to decide what is the best conflict resolution technique to use to resolve the conflict. This needs immediate attention.

Five approaches to resolving conflicts in the workplace

In managing conflict, Thomas (1976) identified five different conflict handling styles used by managers and that they tended to adopt one style rather than another. Successful conflict resolution is based on an accurate and thorough understanding of the actual conflict. There are five well-established approaches to conflict resolution (See also Figure 4, p117).

1. Avoidance

- One of most common responses to conflict is refusing to acknowledge that the problem exists, whilst minimising involvement with the other person. It suggests low concern by the manager for their own and the other person's needs.
- This can be a satisfactory approach if the conflict is relatively unimportant or if there is a need for a cooling off period before the conflict is tackled by management action.
- Often found where a manager lacks confidence.
- Conflicts are often ignored in the hope that they will go away.
- Hoping staff will resolve the conflict themselves.

- Problems are never resolved.
- Risk of being seen as a manager not willing to get involved in resolving conflict.

2. Accommodating (or smoothing)

- This can involve apologising and conceding the issue to the other person regardless of the rights and wrongs of the matter, and goes with low concern with one's own and high concern for the other's needs. Here differences are played down and a façade of harmony erected. This can be a satisfactory approach where the relationship between the parties in conflict needs to be preserved above all else.
- Where one person values their relationship and is less concerned about the eventual outcome of the conflict.
- Submission/giving away.
- Happy to give way on issues that do not affect the service.
- Relationship maintained at the expense of the manager.
- Can risk other staff taking advantage of manager.

3. Competing

- This style tends to be adopted when the manager demands an apology from the other person and is concerned with the redress of perceived 'wrongs', and tends to be adopted by the manager who has high concern for their own needs and low concern for the other person's needs. It tends to be adopted by one party using their authority or position.
- When the outcome is important to you, rather than your relationship with the person.
- Compliance, whether the other person agrees or not.
- Use of power to achieve compliance where there is a need to defend correct position and the relationship is unimportant.
- Can build up hostility and resentment.

4. Compromising

- This is bargaining with the other person until a compromise is reached. The conflict is resolved by each person giving up something in order to meet halfway.
- This is a strategy preferred by many managers.
- Where you are concerned to maintain your relationship with the other person.
- You are happy with a partial solution and go part way to achieving your goal.
- Looking for deals and tradeoffs to find satisfactory or acceptable solutions.
- Works well with mature individuals who accept compromise.
- Risks that other person may play a game of using relationships to demand more than manager can give and s/he looses out.
- This runs the risk that neither party are happy and both feel they have lost and relationships are soured between them.

5. Collaboration

- This approach treats the conflict as a problem that the parties need to solve together. This is usually the best approach, but needs time and the parties to believe in the approach and have the skills to work collaboratively for a consensus solution.

- Joint problem-solving and searching for mutual solutions.
- Both parties comfortable with others' viewpoints.
- Confronting differences and sharing ideas and information.
- Finding situations where all can win.
- May not be useful if decisive action needed.
- Requires a manager who is assertive and co-operative.

These methods are all used by managers and all approaches have their place in the manager's repertoire. It is important to know how and when to use all the options available.

Figure 4 The five approaches to conflict resolution

	Low co-operation	**High co-operation**
High assertion	Competing	Collaborating
	Compromising	
Low assertion	Avoiding	Accommodating

Adapted with permission from Thomas KW (1976) *Conflict and Conflict Management*.

The most important thing is to resolve conflict whilst maintaining your relationship with the members of staff concerned. If you fail to respond to conflict by avoiding or ignoring it and do not engage with the issues that have led to the conflict, this can build up resentment in staff members. Because of the difficulties concerned with managing conflict and disagreements, many frontline managers prefer to avoid or ignore situations and this can lead to more serious consequences (Coulshed & Mullender, 2006).

Moving from informal to formal procedures

Where possible conflicts and disagreements should be resolved informally. This is particularly desirable where there are close working relationships between the frontline manager and members of staff. Where this is not possible, those in conflict need to be made aware that if a satisfactory resolution cannot be achieved through informal management processes then formal disciplinary procedures may need to be used. These include:

- the frontline manager instituting a formal disciplinary procedure
- the staff member taking out a formal grievance
- the staff member making a formal complaint.

Responding to grievance

A grievance is an expression of dissatisfaction by a member of staff or group of staff about a situation where there are problems or complaints that staff want to raise with their employers.

Again as in the disciplinary procedures discussed in the chapter on poor performance, the ACAS guidance (2003) should be consulted as there are statutory procedures that need to be followed to ensure that any action complies with the requirements set out in the Employment Act (2002).

The grievance procedure

- **Step 1:** the member of staff must set out their grievance in writing and send this statement to their employer.
- **Step 2:** the employer must invite the member of staff to a meeting to discuss the grievance. After the meeting, the employer must inform the member of staff of their response to the grievance and notify them in writing of the right of appeal against the decision if they are not satisfied with it.
- **Step 3:** This is the appeal stage and the member of staff has the right to appeal and attend a further meeting, which should ideally be dealt with by a more senior manager. After this meeting the employer must inform the member of staff of their final decision.

At all meetings the member of staff has the right to be accompanied.

Preventing conflicts in the workplace

We have seen that there are different responses to conflict. Some can be avoided, others need monitoring and others require intervention. There are also some general points about preventing conflicts that are worth remembering.

- Recognise and accept that there will be differences between individuals and groups, in terms of attitudes, values, perceptions, expectations and needs.
- Not feeling defensive if others disagree with your ideas.
- Not automatically assuming that you are right and they are wrong.
- Learn to listen to others and what they are really saying.
- Give yourself time to get to know your staff well so that you understand their values and beliefs.
- Try to ensure that both you and your staff learn from conflict situations that have been resolved.

General principles for managing conflicts

Important stages in any process

The sign of a good frontline manager is the ability to resolve conflicts. Managers should know when to take action and when situations are likely to resolve themselves. If you decide not to act this should be a positive decision, not avoidance. It is important to be able to recognise that every situation is unique and make a judgement whether to ignore the conflict, smooth it over or actively confront it accordingly.

Your staff will judge you on how well you manage conflicts and one of the common criticisms made by social workers interviewed was that their managers avoided getting involved in conflict situations and this undermined team morale. When managing conflict, ensure that you make decisions based on a sound understanding of the issues, not on impulse.

Separating the person from the problem/valuing the other person

When you communicate with the other person be careful to communicate that although you disagree with their views, ideas or action, you respect them as a person. You are communicating 'I respect you as a person but disagree with your ideas or actions'. Communication can be non-verbal without words like these being spoken to convey respect for the other person.

The expression of an opposing view is an opportunity to clarify your own. You are acknowledging that there are several ways of looking at a problem.

Problem-solving

This is one step on from recognising different viewpoints and involves combining opposing viewpoints to create better solutions. This is not always realistic but you should always have the possibility in mind. It can take time for solutions to emerge from conflict. Because of our anxiety to resolve conflict quickly there is a tendency to shut it down rather than creating space to find lasting solutions. Seeking a win–win outcome in preference to one with winners and losers (Coulshed & Mullender, 2006).

The stages in resolving a conflict

1. Define the conflict

This involves describing the conflict from each party's perspective and getting a joint definition of the conflict. Try to get to the real source of the problem, recognising the different viewpoints. Enable open discussion to take place, avoiding siding with one or other parties or favouring the more senior person.

2. Actively listen

Encourage the parties to really hear what each is saying and to avoid their feelings and opinions obscuring this. Move on to establishing what the parties agree on and the behaviours they find acceptable and unacceptable. Establish a consensus on possible solutions and agree what each party needs to do to resolve the conflict.

3. Motivation towards resolving the conflict

Look for the gains and losses involved in conflict so that it helps the parties to recognise the benefits of finding a solution. Provide direction particularly where parties are stuck or the problems seem insoluble.

4. Reaching agreement

This should state the agreement reached and how the parties will act differently in the future and any further steps to improve co-operation. Make a clear statement of the agreement so that all the parities receive and understand the same message.

Conclusions

Conflicts and disagreements are common in all organisations. Many will be of a minor nature and are resolved without management intervention. More serious conflicts may need the use of methods that ideally lead to satisfactory resolution for all parties. Being able to assess the seriousness of a conflict or disagreement and knowing when to intervene is a skill that frontline managers need to learn. Not all conflicts should be viewed negatively as they can reveal fundamental questions about issues, particularly those related to service values, policies and practices. Adopting approaches that prevent serious conflicts that have the potential to damage a service, its service users and staff, should be developed by frontline managers.

References

ACAS (2003) *Code of Practice 1, Disciplinary and Grievance Procedures*. London: ACAS. Available from: www.acas.org.uk.

Coulshed V & Mullender A (2006) *Management in Social Work* (third edition). Basingstoke: Palgrave Macmillan.

Goodwin C & Griffiths DB (2007) *The Conflict Survival Kit. Tools for resolving conflict at work.* New Jersey: Pearson Education.

Thomas KW (1976) Conflict and Conflict Management. In: M Dunnette (Ed) *The Handbook of Industrial and Organisational Psychology*. Chicago: Rand McNally.

Managing at the Front Line: Managing diversity

Key points

- **A diverse workforce can more accurately reflect the community served by social care services.**
- **Organisations that promote diversity in their workforce are seen to have many advantages in attracting, motivating and retaining staff.**
- **Familiarity with the main legislation can help frontline managers to develop good diversity employment practices.**
- **Frontline managers can make an important contribution by their personal commitment to diversity policies and practices in their organisations.**

Key words

Equality; different cultures; legal dimensions; codes of practice; leadership; influencing policies.

Social care organisations that are committed to creating a diverse workforce are ensuring that the workforce is as diverse as the people served by it. To ensure that service users receive the best possible service it is important that they are supported by staff who understand and reflect the communities they serve. Managing social care services means supporting the delivery of services to a diverse population of service users living in communities that are increasingly diverse in terms of race, ethnicity, religious and political beliefs, and where people with disabilities are living longer and a growing percentage of the population is over 65 (CSCI, 2006). As a frontline manager you have the potential to make a difference in your service through your personal commitment to valuing diversity and playing an important part in championing the positive development of diversity in social care services.

In managing your service you need to address the main strands of equality practice including, age, gender, gender identity, sexual orientation, race, disability and religion or belief, as these are seen as essential considerations in the delivery of effective social care services. Organisations that value the differences a diverse workforce brings will encourage new ideas and perspectives and as a result will deliver more effective services. For this to happen services need to have policies that are developed against the statutory requirements and set out in equality and diversity policies, with workforce diversity monitored and reported on a regular basis to senior managers.

It is argued that the needs of service users are directly related to workforce developments and that the workforce should reflect the diverse nature of the communities it serves (Drennan *et al*, 2004), and that organisations should develop cultures in which diversity is acknowledged, valued and effectively managed, and that this is likely to increase access to quality services at the front line (Friday & Friday, 2003). The structures and cultures of services can mean that the contribution of staff in terms of their background and diversity are not recognised and that staff from different cultures are not valued for what they bring to an organisation (Hafford-Letchfield, 2006).

Benefits of an equality and diversity promotion

The Commission for Equality and Human Rights (www.equalityhumanrights.com) states that there are considerable advantages in promoting race, disability and gender equality in organisations. For example, it can help:

- achieve a more representative workforce
- attract and keep able staff
- improve staff morale and productivity
- improve staff management
- identify and develop good practice
- identify discriminatory outcomes for any group of employees
- avoid claims for unlawful discrimination.

This quote from a manager of a learning disabilities service captures some of the benefits of a diverse workforce.

> *'We are serving a community which has a rich ethnic and cultural diversity. This means that our service users and staff come from that community. We have found it an immense advantage to the service in many ways, including an economic benefit in terms of recruitment, a practical benefit in terms of the range of perspectives that our staff offer and a cultural benefit in terms of the creativity that staff bring to the service. Our service is founded on the principle of valuing difference so we don't want a workforce where everyone thinks and behaves the same, rather we are seeking a harmony that unifies our diversity.'*

Legal dimensions

From a legal perspective you need to be familiar with the main legislation that impacts on your service as a 'public body'. This includes: the Sex Discrimination Act (1986); the Race Relations Act (1976) and the Amendment Act (2001); the Human Rights Act (1998); and the Disability Discrimination Act (1995). In addition, there is the new Duty on Age (2006) and the Gender Duty (2007). All these acts and duties place equality at the centre of employment practice and require you to:

- eliminate unlawful racial and disability discrimination
- promote equality of opportunity
- promote positive attitudes to disability
- promote good race relations between people of different ethnic groups
- consider how to eliminate harassment
- encourage the participation of disabled people in public life
- take into account gender, including gender identity, issues
- avoid age discrimination.
  (CSCI, 2006)

Reference to diversity is also made in the Standards for Qualifying Social Workers (www.skillsforcare.org.uk) and in challenging discrimination in the General Social Care Council's Codes of Practice for employers of social care workers (www.gscc.org.uk). Similarly, CQC (www.cqc.org.uk) is incorporating equalities into its review of the National Minimum Standards for social care services.

Leadership and diversity

As a frontline manager you are in a powerful position to encourage and enable members of staff to develop their leadership potential. We have seen in Chapter 4 that the notion of distributed leadership and the promoting of leadership in others has much to commend it in enabling organisations to encourage those groups who have traditionally been excluded from positions of responsibility due to practices that saw leadership as a predominantly male attribute (Rogers & Reynolds, 2003).

Evidence from research in social services departments suggests that, although women make up the majority of the workforce, there tends to be occupational segregation that results in women being found predominantly in lower paid jobs. In their research into working in social services, Balloch *et al* (1995) found considerable occupational segregation by gender. Although women comprised the overwhelming majority of the social care workforce (86–95%), they comprised only 60–70% of those in management roles. Women were significantly under-represented in senior posts, due in part to gender differences in their work histories stemming from traditional family roles. Women were also less qualified, older and had shorter full-time service then men, which in turn reduced their chances of promotion to senior posts. Hourly pay was less for women than men within each job type, indicating that women were employed at lower grades than men within occupational groups. A later study similarly found that organisations were failing to capitalise on women's experience and their aspirations towards career progression (NISW, 1998).

In terms of black staff, 27% had experienced racism from colleagues or managers, with those affected not receiving adequate help and support from their organisation. Black and Asian staff felt that there was insufficient attention paid to equal opportunities and that to make progress in their careers they had to be much better than white colleagues (Balloch *et al*, 1995). A 2001 study (SSI/ADSS, 2002) found that only a tiny minority of women managers were of non-white ethnic origin and the majority of those were in junior management posts and based in London. A separate study (Bhavani & Coyle, 2000) reported that women from black and ethnic minorities felt frustrated about their chances of career progression and that their capabilities were not recognised. This was further confirmed in the recent report from the Government Equalities Office (2008).

In their discussion of distributed leadership, Rogers and Reynolds argue that the focus on extending opportunities should be extended to others who have traditionally had less opportunity to take on leadership roles. This includes people with disabilities and gay and lesbian people and that the reasons for doing this include, *'better human relations practice, assumptions about equity and fairness, a better reflection of the population served by the organisation and the potential loss to the organisation of the creativity of people from diverse social, racial and cultural backgrounds'* (2003: 77). This latter view is reinforced by the argument that when people from diverse backgrounds hold power it can add value to an organisation, but that it requires an organisational development strategy and change programme for this to succeed (Bhavnani & Coyle, 2000).

Rogers and Reynolds (2003) put forward a series of questions that can be helpful in assessing diversity and opportunity in services.

- Who benefits from the current working arrangements, in terms of services delivered, in the organisation and in my team?
- Which group dominates this social arrangement, for example, committees, working groups, departments?
- Who defines the way things are structured in the organisation?
- How do practices, process and systems promote or restrict inclusiveness, for example, what characterises my profession, my day-to-day work, my relationship with senior and subordinate colleagues?
- How does language commonly shape attitudes and actions and where does it come from?

These can be useful questions to start you thinking about the design and structure of your organisation and the practices that are promoted and how much it can respond to the full range of human needs and potential.

Actions that you can take or influence

As a frontline manager you need to be familiar with the main legislation and ensure that you comply with the different laws. Similarly you need to ensure that staff members are committed to understanding issues of diversity as this impacts on their work. As staff increasingly work with diverse groups with individual differences it is important that they develop their knowledge and understanding of the impact that culture has on service users who are seeking help, and whether they feel there is stigma attached to certain conditions (Malcolm, 2007).

There is a number of specific actions you can take as a frontline manager to support the diversity in your organisation, alongside the wider organisational HR role in monitoring of groups that are under-represented in the organisation and the strategic actions that demonstrate commitment from senior managers. The following are examples of positive actions you can take.

- Have a good understanding of the legal framework as it applies to public bodies and familiarise yourself with the guidance provided to employers and employees found on the Commission for Equality and Human Rights website (www.equalityhumanrights.com).
- Adopt an outreach approach to attracting staff from under-represented groups by ensuring that jobs are advertised in media that serve particular groups of staff, with advertisements welcoming applications from under-represented groups.
- Raise the profile of the organisation and its credibility as an employer through contact with, for example, black and minority ethnic organisations or those for disabled people.
- Establish mentoring schemes and support staff from black and ethnic minority, disability and women's groups, with specific courses or schemes for those who wish to progress their careers.
- Support staff training opportunities to ensure that they understand what diversity means to their service and that this is incorporated into their day-to-day practice.
- Demonstrate your own commitment to promoting and valuing equality and diversity through your own behaviour and challenging inappropriate behaviour or incidents of discrimination.
- An understanding of diversity issues means you are more likely to confidently tackle poor performance, particularly in areas where the dynamics of oppression make it more difficult to deal with issues and where you could be accused of oppressing a member of staff. (Coulshed & Mullender, 2006).

Some of these initiatives will require organisation-wide strategies, but you can nevertheless play an important part in ensuring that equality and diversity are central to the work of your team and that this is recognised throughout the organisation. It is easy for organisations to trumpet their equal opportunities policy without a real commitment to improving practice (Coulshed & Mullender, 2006). What these actions suggest is that as a frontline manager you have an important culture shaping role in valuing and managing diversity in the workplace and can create the conditions in the service that you manage to foster and value cultural differences, and that creates a positive work culture.

Conclusions

It is increasingly recognised that a diverse workforce offers considerable advantages to the effectiveness of an organisation in reflecting the communities served and providing more creative responses to service users' needs. There are a number of legal requirements, codes of practice and national standards that underpin the development of more diversity in social care services. Frontline managers can play an important role in influencing the culture, policies and practices that increase diversity, which in turn contributes to effective service delivery and practice.

References

Balloch S, Andrew T, Ginn J, McLean J, Pahl J & Williams J (1995) *Working in Social Services*. London: NISW.

Bhavnani R & Coyle A (2000) Black and ethnic minority women managers in the UK – continuity or change? In: MJ Davidson & RJ Burke (Eds) *Women in Management: Current issues, Volume II*. London: Sage.

Coulshed V & Mullender A (2006) *Management in Social Work* (third edition). Basingstoke: Palgrave Macmillan.

CSCI (2006) *Equality and Diversity Strategy*. London: CSCI.

Drennan V, Levenson R, Goodman C & Evans C (2004) Developing the health and social care workforce. *British Journal of Health Care Management* **9** (1).

Friday E & Friday S (2003) Managing diversity using a strategic planned change approach. *Journal of Management Development* **22** (10).

Government Equalities Office (2008) *Routes to Power, Summary of research*. London: The Stationery Office.

Hafford-Letchfield T (2006) *Management and Organisations in Social Work*. Exeter: Learning Matters.

Malcolm B (2007) Managing diversity in social services settings. In: J Aldgate, L Healy, B Malcolm, B Pine, W Rose & J Seden (Eds) *Enhancing Social Work Management.* London: Jessica Kingsley Publishers.

NISW (1998) *The Social Services Workforce in Transition. Research summary 13.* London: NISW.

Rogers A & Reynolds J (2003) Leadership and vision. In: J Seden & J Reynolds (Eds) *Managing Care in Practice*. London: Routledge.

Social Services Inspectorate/Association of Directors of Social Services (2002) *Women Rising? Achieving equality and diversity in leadership: An SSI/ADSS sponsored study of women in top positions*. Bristol: Social Services Inspectorate.

Section 3

Chapter 15

Managing at the Front Line: Managing change

Key points

- Change is a constant feature of all organisations, ranging from incremental improvements to large-scale restructuring.
- It is important to anticipate the need for change and continually improve a service.
- Understanding the impact of change on individuals is important when you are managing the change process.
- Resistance to change should be seen as an opportunity to explore the impact on staff affected by the change.
- Different styles of managing change can be used ranging from the participative to directive depending on the urgency and scale of change demanded.

Key words

Incremental; strategic; change processes; resistance; ownership; leadership.

One of the central tasks of the frontline manager is to continually improve the performance of the service they manage in order to provide a more effective service for users. As a frontline manager close to practice you have an important role in initiating change for service improvement. However, you also play a role in responding to imposed change in the service triggered by the need to respond to new policy initiatives, or by opportunities presented by forces external to the service.

The ability to respond to changing demand and expectation through developing and shaping the service is a crucial aspect of management effectiveness, but one that requires an understanding of the wide range of issues involved in the processes of change. Although organisations experience periods of continuity and stability, these are tending to become shorter as services become more sensitive and responsive to changes and pressures from a wide range of stakeholders. Organisations are now more likely to reject the notion of fixed structures and systems, and support an approach that values continuous adjustments to systems as part of managerial improvements in services.

It is important that frontline managers generate new ideas and put them into effect as part of developing the organisation's ability to respond to the changing needs of users and carers and demands on the service. A desirable goal is a service that is flexible and responsive and able to adjust to changes in the environment on an incremental basis. This approach means that managers anticipate emerging and future needs and organise their resources around them, rather than being unresponsive and ignoring change, until it is forced on them.

Social care organisations are healthier where they focus clearly on the needs of users with a culture where staff are keen to engage constructively with change agendas. In contrast, closed and controlling organisations absorb new agendas with little change in practice or at best superficial changes. The reports of inquiries into learning difficulties services (CSCI, 2007) demonstrate how these services failed to change in spite of overwhelming evidence that they were no longer providing acceptable services. The examples of resistance to change in these services demonstrated the power of organisational cultures to resist change in spite of overwhelming evidence of the need for improvement.

Thinking strategically and developing a responsive service

A key aspect of your role is an awareness of the broad thrust of government policy, how this is interpreted in your own organisation, and how this will impact on practice. You should also understand the part that your management plays in supporting the implementation of policies through change processes. This awareness has a number of features:

- understanding the strategic issues impacting on your organisation and how the activities of your staff are related to these and what needs to change at the front line of the service
- identifying the scope of your team's activities within overall organisational strategies and matching these activities to the environment in which you operate
- matching activities to the resources available to the team in terms of core competences such as knowledge and skills in order to deliver an effective service.

An important element of good management is understanding the changes taking place in the environment of your service and responding to these through putting in place the day-to-day changes where you make improvements in work practices to improve efficiency or support team members on how they can introduce changes in their practice. These changes could all be considered incremental changes as you are continually adapting the service and its response to changes in the internal or external environment that is impacting on the service.

As a frontline manager you are in a unique position to develop an understanding of the changing needs of service users and carers. This means you should be continually thinking about what changes in the service could meet these changing needs. Your approach should be one where you are able to respond rapidly and positively to change opportunities. This means not waiting until senior managers initiate change but being responsive to changing expectations and demands where you are in the best position to introduce incremental changes based on feedback from the front line of practice. This still means that you many need the sponsorship of your manager or other senior managers to ensure that you have the authority to introduce change.

When large-scale changes are instituted at senior management level, your job as a frontline manager is to interpret what it means for your team and how the change can be introduced and maintained without damaging the service and its users. Senior managers by the very nature of their role and position in an organisation tend to see change in terms of ideas that someone else has to translate into practice, with less concern about the messy details of implementation. That is your job! But don't just be someone who reacts to the change agenda; seek to control it so that you can understand its purpose and what it means for your team and the service they provide. Can it be integrated into the current ways of working, or will it need significant change to ways of working?

The power of culture

Whatever the scale of change, it is important to understand the power of organisational culture and its influence on the change process (Scragg, 2007). Culture has a powerful constraint on organisational change. Small-scale incremental changes may not challenge established routines and working practices, or in other words, 'the way we do things around here'. Changes at this level are contained within the existing ways of working and it is less likely that staff will resist change. Whereas large-scale change may mean radical change and is more likely to be questioned and resisted, at least by some staff, and therefore more difficult to manage. At whatever level change is considered, it is important to understand the power of culture and its influence on change. Without an understanding of culture it is difficult to understand why, what seem to the outsider and the manager trying to manage the change process, logical and reasonable attempts to improve a service fail, or repeated attempts to change practice proceed slowly and require many attempts to reform practice.

Incremental change

For most organisations incremental change is likely to be the main way that a service changes and adapts through continually responding to changes in its environment, through adaptations to practice following the introduction of new policies or where there are changes in expectations or new areas of need identified. This is the gradual change which makes the most sense because it is change within the preferred way of doing things and does not threaten established forms of practice and routines. It is also realistic as no organisation can function effectively if it is continually undergoing major reorganisation with all the consequences of turbulence, risks to staff morale and loss of focus on the needs of service users. It also builds on the existing skills, routines and beliefs of staff so that changes are likely to win their commitment.

But there are risks in incremental change as an organisation can become complacent and fail to challenge or question itself and fail to react to clear changes and demands taking place around it. Numerous examples of this can be found in the reports of inquiries into services for older people and learning disabilities services, which failed to change in spite of repeated concerns about the declining quality of services or outmoded practices and abuse (Scragg, 2007).

This is described as *strategic drift* where, over time, services gradually become less and less relevant to the needs and expectations of those that they serve. It is in these situations that an organisation is required to challenge their core assumptions about the services they provide. We are beginning to see examples of this where services have to critically examine what they provide in a era where service users with individual budgets are making choices that threaten the survival of some provider services. Examples of drift are seen in local authority day services for people with disabilities where service users are being given the freedom to choose whether to continue using a service that fails to meet their needs, and where controlling their own budget provides more creative options.

Transformational change

Where large-scale system level change is called for, this is described as transformational change, as it often demands radical reform, including changes to the mission and values of the organisation, and considerable restructuring, including new roles and functions (Johnson & Scholes, 2002). Where change is transformational it means that change cannot be handled within the existing organisational systems and routines, and a change in the taken-for-granted assumptions about the way of doing things. This level of change can be planned or forced, dependent on the urgency of the need for change. An example would be where a service had failed to respond to change (strategic drift) and where forced transformational change was necessary to save a failing organisation. The example quoted above of the residential service for people with learning disabilities demonstrates the pressure for transformational change that was forced on the organisation (CSCI, 2007).

It is important to be aware of the risks involved in transformational change when the focus on the needs of service users can be lost temporarily as staff become preoccupied with structural change and new roles and relationships. Major structural change needs to have overwhelming arguments in its favour (Gulliver *et al*, 2002) and the risks need to be borne in mind when you are tasked with implementing transformational change and the potential negative consequences for service users. Of course, change on this scale may be essential where services are failing to meet the needs of service users or have become increasingly irrelevant to the people they were established to serve.

Responses to change

In the past, change was seen as an event that interrupted the normal functioning of an organisation. This is now changing as people become more skilled in managing change and more positive about initiating change to improve a service. How staff respond to the opportunities will depend on their disposition to the intended change. Some may be keen on innovation and trying new ways of doing things, whilst others may be well disposed to change if they see benefits to themselves and the service. Others may be resistant, or at least sceptical, until they understand the change more fully and how it will affect them personally.

How staff respond to change will depend on a number of factors which affect them. The following statements were made by frontline managers from group care services who had experience of implementing significant change.

- *'Will it mean a change to established ways of working?'*
- *'Will it change social relationships at work?'*
- *'Will it threaten job security or job prospects?'*
- *'Will it lead to loss of status or self-esteem as a result of enforced changes in role?'*
- *'Will it mean staff are de-skilled until they learn new methods and achieve new performance standards?'*
- *'Will it lead to an increase in workload on a temporary or long-term basis?'*
- *'Will training be provided to help develop the new skills needed?'*

With so many factors potentially affecting people involved in change it is understandable that some staff may adopt defensive or negative attitudes and resist proposed changes. Your main task is to work with your staff to enable them to make sense of the need for change, what is required of them and what support they will need.

Good practice in managing change

There are many approaches to the management of change, but all tend to focus on a number of essential elements that you need to incorporate into your practice. That said, change is rarely straightforward, and even the best prepared change process will need to acknowledge that people react to change in different ways. Change rarely takes a linear path from introduction to implementation. To manage a change process effectively you will need to understand the individual's perspective on change, and the range of responses generated by the introduction of new ways of working.

Your own leadership of the change process

A major factor in the introduction of change is your own behaviour. A participative style of leadership is helpful as this will enable you to discuss the reasons for change fully with staff and encourage their questions about the change (Gilbert, 2005). This will enable staff to understand more fully the reasons for change and will also provide valuable feedback for managers on the potential for success. Similarly, involving staff in the implementation process in order to generate commitment and ownership is also crucial.

Trust in your leadership

All large-scale change means stepping into the unknown to some extent and staff will need to trust you as their frontline manager if you are to work effectively in the 'change agent' role. The success of any change will be dependent on a range of often interrelated, complex factors (Gilbert, 2005). It is critical that staff trust you to communicate openly with them, to be sensitive to the consequences of your actions and to be prepared to change those actions based on the feedback you receive.

Clarifying what is involved in the change

An important early stage in the management of change is first to define clearly the purpose of change – what is the objective of change and what are the reasons for it? (the why). Second, what is the nature of the change – what will be involved in the change and the proposed effects of change? (the what), and finally what are likely to be the main effects of the change on the organisation and the staff?

Resistance to change

Resistance to change is not in itself a bad thing. It can be useful in forcing those pushing for change to examine what is being expected of staff and reconsidering whether the rush to implement change has been well thought out and if change is being introduced for its own sake rather than for the benefit of the service and its users.

Resistance to change can have a number of possible sources, including the reasons for change not being fully explained, the status quo being seen as satisfactory and change not being justified, staff not feeling engaged with the change process, the impact on work groups, splitting up long-standing professional partnerships, and the affect on the shift system or working hours. Resistance may also be related to the comfort that staff feel working in well-established ways that are threatened by change. Concern about what the future will look like following change is a large element in resistance. It is important to listen to people who resist change and to understand what fears about the consequences of change underlie their opposition.

Some ways of working with resistance

- Explain fully why the change is necessary, and what problems it is designed to overcome. This provides staff with a contextual understanding, explaining the internal/external factors that are driving the need for change.
- Encourage staff to participate in the change and engage with the change process. This will provide an opportunity for those most closely affected by change to suggest ways of improving the planning process and having some ownership of the change process.
- Provide staff with the opportunity to voice their objections to the change. This will enable misunderstandings and misconceptions to be discussed and resolved and reveal doubts about the change. It can be useful where plans for change have not taken into account implementation issues and the greater understanding of frontline staff.

Recognise that there are likely to be some members of staff who are resistant to change no matter how much the change will benefit the service and its users and who will question the need for any change, and also challenge your legitimacy to introduce change in the service. Work with numerous managers who have expressed their frustration in dealing with such staff has indicated that it is better, after reasonable attempts have been made to inform and persuade such staff of the need for change, to work with those staff who express commitment and develop a coalition of staff who are willing to examine the change and how it can be implemented.

Working with resistance may also help you to identify where the change is not going to work and needs modification. Because change often fails at the implementation stage, it is a chance to influence the shape of change so that it is more effective and has a greater chance of implementation. Yet it is implementation that is one of the main causes of failure in change. It is one thing to have a strategic reason for change, it is another to convince managers and staff at the front line that the change is necessary.

Those with influence and power

An important factor in introducing change is identifying those staff with informal power and influence that are likely to be critical to the change process. In many examples of change in group care services, long-serving members of staff who have influence among colleagues can present a major stumbling block to the manager who is tasked with introducing changes, or identifies the need to make changes to improve the service. Don't waste valuable time trying to persuade staff who may take delight in frustrating your efforts as a demonstration of their power and influence. Such staff need to be given clear and unambiguous instructions and the requirement to comply with new ways of working (Hudson, 2002).

Some staff may also be skilled at absorbing and diluting new ideas in accordance with their own norms, which means that change is only superficially adopted and momentum is lost. There is behavioural compliance without a real commitment to change so that change is superficial in its implementation. This form of covert resistance is much more difficult to manage and risks undermining change prospects.

Identifying forces for and against change

The force field analysis is a well-known tool for assessing the forces driving and resisting change and is based on the original idea by the social scientist Kurt Lewin (1957). It is a diagrammatic way of mapping the forces that are keeping a particular situation in the current state of equilibrium between the forces pushing for change (driving forces) and the forces resisting change (restraining forces). The force field analysis can help provide a picture of the receptivity to change and identify those factors that can help you implement change. Although you will want to maintain the important driving forces, paradoxically trying to increase them tends, in turn, to increase the restraining forces. Your focus should be on working with staff to identify their concerns rather than trying to push through change, which may only cause more resistance. As Martin and Henderson (2001) suggest, you may decide that it is not possible to implement the proposed change that could lead you to question whether your time and resources would be better served on a development with a greater prospect of success.

Figure 5 Force field model diagram

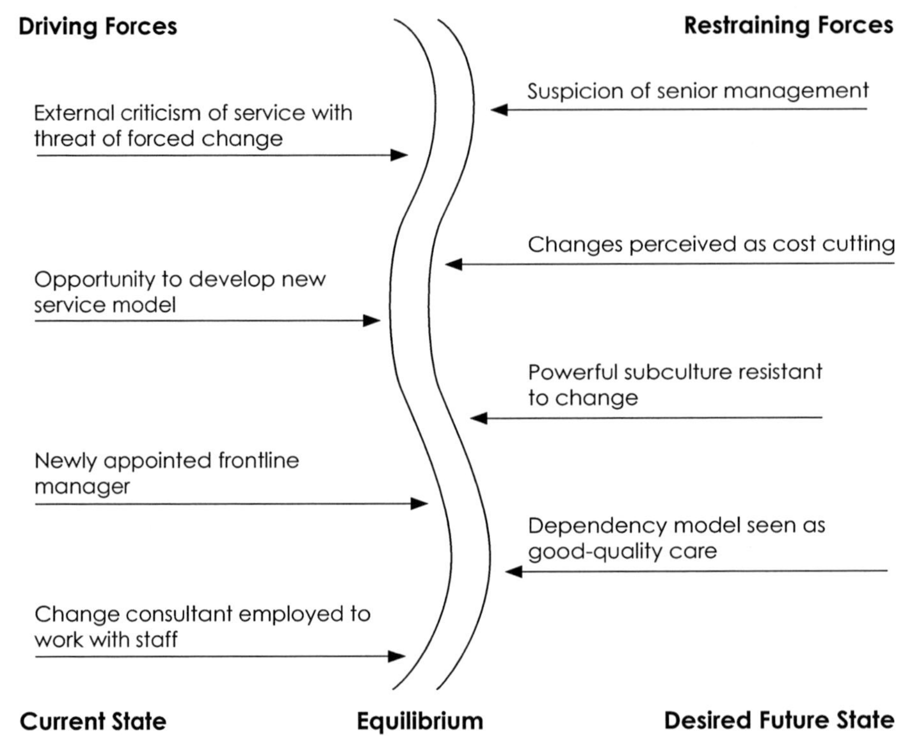

This example of a force field analysis illustrates some of the driving and restraining forces present in a service for disabled people that had been criticised for its outdated practices, and which was attempting to develop a new model of care based on closure of the institutional service, and its replacement by a more individualised community-based service.

Styles of managing change

When you are leading the change process it is helpful to consider the most appropriate styles of management to adopt to achieve the aims of the change process, as this will differ according to the organisational context and the degree of pressure for change (Johnson

& Scholes, 2002). The following styles describe a range of approaches, with different combinations of style adopted according to organisational context and the degree and urgency of the change required.

- **Education and communication:** are used when staff need to learn new skills that are crucial for working in new ways. Communicating the vision through meeting with those working at the front line, testing out the vision and reinforcing it is an essential role for change managers.
- **Collaboration and participation:** are used as a means of engaging with those affected by the change, asking them to help design the process and thereby increasing ownership and commitment to the change.
- **Intervention:** is used when a change agent or others are given delegated responsibility to work with staff on particular aspects of the change process, including, idea generation, detailed planning or the development of rationales for change. Again, involvement is likely to lead to greater commitment to change.
- **Direction:** is used when those leading the service use their authority to set the direction, means and speed of change. Closely associated with this style is *coercion*, where change is imposed through the use of managerial power. These styles may be used when an organisation is facing a crisis and survival is heavily dependent on change.

Whichever methods are used, and most change processes use a combination of methods, there are advantages and disadvantages. A management style based on education, communication, collaboration and participation can be time consuming, although they are likely to increase ownership and commitment to change. Intervention, direction and coercion have advantages of speed, but risk perceptions of manipulation or, at worst, non-acceptance by staff. To a great extent, the urgency, scale and readiness for change will dictate the style adopted by those leading change, and any of the styles may be needed depending on the wider context within which change is taking place. Planned incremental change will favour a more participative style, whereas forced transformational change will favour a more directive approach.

Some of the most difficult aspects of the process of change focus on everyday routines, which are part of organisational life, and in turn this challenges the taken-for-granted assumptions and deep-seated beliefs that are the central elements of an organisation's culture. Pascale (1990) argues that it is often more effective to involve staff in the activities of changing, including changing routines, rather trying to change behaviour through persuasion and attempts to change taken-for-granted assumptions. This has implications for the style of change employed and that managers need to adopt different styles at different stages of the change process.

Key activities during the change process

There are a number of activities that are important when managing the change process and the following points are important to remember as they can make the implementation more successful.

- Explain the purpose of the change and why it is necessary. Provide staff with an understanding of the reasons for the change.
- Communicate a clear sense of direction – expressed ideally as a 'vision' – which staff can relate to the aims of the service.
- Involve those staff affected by change at an early stage and take account of their ideas. Provide opportunities for staff to 'own' the change, although remember you have overall responsibility for the strategic direction of the change.
- Be aware of the power of culture and its influence on how staff see the proposed changes. Norms and social relationships of the staff groups are often affected by change and these will also influence how staff view the change.
- Listen to staff and their concerns, acknowledge their fears and work to turn threats into opportunities. Provide staff with space to question the change and state any concerns or objections as this will enable them to feel more committed and also put forward suggestions for improving the implementation.
- Identify gaps in skills and arrange developmental opportunities to help staff work in new ways, with access to training where new skills are needed.
- Emphasise the qualitative aspects of change, such as building commitment and trust as confidence in those leading change is vital.
- Recognise that the participative approach to change can be time consuming and that real progress in changing takes time.
- Ensure that staff do not have excessive demands on them during the change implementation phase and that they have the resources needed for the new ways of working.
- Avoid over-organising as the impact of change can rarely be anticipated.
- Identify some short-term successes that are recognised and create further commitment.

Conclusions

Change is a feature of all organisations, whether it is the day-to-day improvements in practice or the major changes to structures and systems. Frontline managers need to understand how change impacts on individuals so that they can sensitively manage the change process. Working with staff who are sceptical about change demands strong leadership skills, with a willingness to engage with staff and increase ownership of change. Being familiar with different styles of managing change and the actions that can be influential in implementation should be understood.

References

Commission for Social Care Inspection and Health Care Commission (2007) *Joint Investigation into the Provision of Services for People with Learning Disabilities at Cornwall Partnership NHS Trust*. London: CSCI.

Johnson G & Scholes K (2002) *Exploring Corporate Strategy* (sixth edition). Harlow: Pearson Education.

Gilbert P (2005) *Leadership. Being effective and remaining human.* Lyme Regis: Russell House Publishing.

Gulliver P, Peck E & Towell D (2002) Evaluation of the integration of health and social services in Somerset: Part 2: lessons for other localities. *Managing Community Care* **10** (3) 33–38.

Hudson M (2002) *Managing with Profit*. London: Directory of Social Change.

Lewin K (1957) *Field Theory in Social Science.* New York: Harper and Row.

Martin V & Henderson E (2001) *Managing in Health and Social Care*. London: Routledge and The Open University.

Pascale R (1990) *Managing on the Edge*. Harmondsworth: Penguin Books.

Scragg T (2007) Organisational cultures and the management of change. In: A Mantel & T Scragg. *Safeguarding Adults in Social Work*. Exeter: Learning Matters.

Managing at the Front Line: Managing your budget

Key points

- **Frontline managers are increasingly responsible for the management of the budget for their service.**
- **It is intended to give frontline managers more control over decisions about expenditure.**
- **Frontline managers should be fully involved in planning the budget for their service.**
- **It is important for frontline managers to understand the language of budgets so that monthly statements can be monitored effectively.**
- **Frontline managers should be supported by their finance departments.**

Key words

Devolved responsibility; budget planning; monitoring monthly statements; support to manage a budget.

Responsibility for budget management is increasingly being devolved to frontline managers. There are a number of reasons for this. First, it links managerial and financial responsibility at the point where resources are deployed. Second, it enables frontline managers to make better decisions on priorities. Third, it facilitates more flexible responses to the needs of service users and, finally, it provides incentives for the cost-effective use of resources.

Additionally, the wider managerial changes introduced since the late 1980s have seen public services come under increasing scrutiny with regard to public spending and have resulted in devolving as many decisions as far down the hierarchy as possible, linked to a parallel move away from centralised management structures. Similarly, services in the independent sector are under pressure to reduce costs as part of market-based competition between providers (Gallop, 2003).

Giving frontline managers the control of the budget for their service is intended to help them to make more effective decisions and manage the budget allocated to their service better.

This process ideally enables the manager to see the relationship between the way resources are deployed and the expenditure under their control. It offers the opportunity for a greater degree of control over the budget management process and enables managers to allocate resources to areas that require support based on frontline information, thereby improving the overall performance of the service.

> *'Since I've been responsible for my budget I've been able to alter the staffing to meet pressures on the home. I've reduced a management post to be able to employ more care staff as the dependency levels of service users is increasing. I didn't have that flexibility before.'*
>
> Frontline manager, group care service

What is a budget?

Budgets express the aims of an organisation in financial terms. They are a statement of the resources that are available to achieve the aims, and are the overall sum of money that a manager has to control. Budgets are normally drawn up for 12 months and are relatively short-term in comparison with the strategic aims of the organisation, which may cover several years. Martin and Henderson (2001) identify a number of purposes for budgets and their contribution to improving the effectiveness of an organisation:

- **planning** – using the aims of the organisation to plan activities in detail
- **communication** – senior managers and others communicating what is expected of frontline managers and staff
- **co-ordination** – the budgeting process encourages managers to communicate with one another and thereby co-ordinate their activities
- **motivation** – challenges managers and staff to achieve the best they can within the resources available to them, although budgets can also be demotivating if staff feel that they are unable to provide a quality service where funding is restricted
- **control** – using monthly budget statements to discover whether activities are being carried out within budget and with remedial action taken to ensure targets are reached
- **assessing performance** – budgets can be used to assess a manager's performance in relation to their budget responsibilities.

The budget cycle

The budget cycle in most social care organisations traditionally begins in the autumn each year (if their financial year ends on the 31 March) as managers consider the needs of the service in relation to existing commitments and any proposed developments. By November, local authorities will have been informed by the government about their revenue support grant for the coming year, and by January, will have agreed their service budgets. This in turn will be communicated to independent sector organisations who are providing services on contract to local authorities.

Re-forecasting

Some organisations re-forecast their budget every six months to take account of the changes in their financial situation. When re-forecasting takes place, the budget is adjusted on the basis of what has happened since the beginning of the budget period to take account of changes in income or expenditure. This can occur, for example, when new contracts are agreed or an anticipated income is less than originally planned.

Methods used to prepare budgets

There are several stages to the budgetary process. A budget is essentially a plan for the coming year that is expressed in financial terms. In preparing the budget, an organisation will examine its planned activities for the year, which are then translated into the financial resources needed to implement them. The traditional way in which resources are allocated in social care services is by a process called 'incremental budgeting'. The current year's budget is the base budget and it is updated (incremented) for inflation and other commitments. The main determinant is what expenditure has been incurred in the current year.

The important questions that this form of budget planning raises are whether items are still required and the amounts are allocated appropriately. The alternative is zero-based budgeting, which ignores the previous year's budget and where every item is assessed for the first time. Incremental budgeting is nevertheless widely used because so many items and activities are for statutory services or are so fundamental to the organisation that they will continue each year. What is important is that the activities are continually reviewed and, where appropriate, zero-based methods used.

Some problems with devolved budgeting

In spite of the increased shift to devolved budget management at the front line, this process can be undermined by a number of problems that can affect the commitment of frontline managers to endorse budget management as identified by Field (1999):

- budget preparation and control is separated from service planning and management
- budget managers have little scope to manage, and in effect can only administer their budgets
- managers who feel unable to shape their budgets can lack the commitment to achieving a budget they have had no hand in preparing
- start–stop spending during the year, culminating in 'spend it or lose it'
- unnecessary involvement of the finance department staff in budget management.

There are four main reasons for this situation according to Field, including:

- the organisation fails to recognise that devolving budgets means a shift in power, with increased discretion at the front line and reduced detailed central control

- budgets continue to be prepared by someone else and are subject to 'cuts', which the manager is expected to absorb
- the financial systems are based on those designed for accountants, and managers find the presentation of information unhelpful
- continued use of incremental budgeting, often with the application of *pro rata* cuts even though there are greater demands on the service, which must be achieved with fewer resources.

It is important that organisations avoid having senior managers imposing budgets, as this can be demotivating to managers and staff and also undermines the organisation's goals. If frontline managers participate in budget planning, it can help them to understand where there needs to be compromises and trade-offs and also why there may need to be budget cuts. The participation of frontline managers in setting targets for the service and the associated budgets should be a motivating activity that helps the organisation achieve its strategic goals (Palmer *et al*, 2002).

Budget planning

This is an important part of the budget holder's responsibility. If you are responsible for the budget then you should be involved in the process of preparing it. Participating in the budget-planning stage will enable you to consider whether your current budget will meet the demands of the coming year's activity and how conditions may change that have financial implications for the service. You may also have identified areas of the service where additional resources are required to achieve particular levels of service and can be supported by budgetary allocations. Participating in this way provides greater ownership of budgets and increases the commitment to managing budgets effectively and meeting agreed targets.

What budget management means for you as a frontline manager

As a 'budget holder' you are given authority to spend funds in order to achieve targets in relation to the overall plan for the service. As the budget holder you are the person responsible for controlling expenditure against one or more budget headings and ensuring that funds are available before making a commitment on expenditure. You also need to identify any expenditure that appears against your budget statement that is in error. You will have an opportunity to monitor your expenditure through the regular monthly statements, which state expenditure being incurred against the budget. This information is increasingly being provided online as organisations develop their computer networks to local offices and other locations.

'We've devolved budgets to individual budget holders and they are in control now with the power to decide what to do with their budget. The decision-making is easier for them when the resources are under their control. The budget information is to enable them to do their job, but we're here to back them up. If they're worried we are on the end of the phone. I see us as part of the team supporting the frontline managers.'

Finance manager, group care services

Monitoring your budget

Once your budget is agreed you will need to monitor it. Each month you receive a statement of your expenditure to date compared to the budget, and usually the current month and year-to-date position. The budget statement is the means by which the organisation communicates information to you about the activity levels and financial consequences of the previous month's activity. This statement provides you with information about the income and expenditure you control, and enables you to compare your budget with actual expenditure and income. Your responsibility as the budget holder will be to 'keep within your budget', although you need to be clear about what your responsibilities are and how much control you have over budget items.

The budget statement will be divided into columns and rows, which reflect the actual and anticipated profile of expenditure for your service throughout the year. The rows will usually be divided into pay and non-pay items, with the pay items usually the greater part. Non-pay items are usually a smaller element of your monthly budget statement, and refer to items such as utilities, supplies and equipment.

Questions to ask about your budget statement

When you look at your budget statement you should look at the overall performance of the budget. Whether it is in surplus or deficit and whether the statement looks right from your understanding of the service. For example, if you have a number of staff vacancies you would expect the staff costs to be under budget. Questions you should ask yourself when you are reading your budget statement could include the following.

- Am I clear about the budget information I have received?
- Is the summary accurate and giving me the correct information?
- Are there any transactions that you did not expect – is this because of miscoding that needs to be corrected?
- Are there any transactions missing? This could be because some income has not been received.
- Is the information outdated?
- Am I clear about the costs I can control and those I cannot?
- What is the extent of my power in relation to your budget?
- Am I held responsible for overspends?
- What are the limits of my authority for expenditure?

What you need to clarify is the extent of your responsibility as a budget holder and what freedom you have to balance expenditure between budget heads and to control spending in terms of staffing (usually your major expense) and the purchase of supplies and services. If you have any questions about the monthly statement or the way your authority incurs expenditure you should contact your finance manager.

Figure 6 Budget example 2007/08 Lyndale House

Month: July	Actual month (£)	Budget month (£)	Variance (£)	Actual YTD (£)	Budget YTD (£)	Variance (£)
Income						
Residential fees	18,000	20,000	(2,000)	77,000	79,000	(2,000)
Total	18,000	20,000	(2,000)	77,000	79,000	(2,000)
Expenditure						
Salaries and wages	9,000	12,000	3,000	37,000	46,000	9,000
Rent and utilities	600	600	0	3,000	1,000	0
Vehicles/tax/fuel	600	600	0	2,400	2,400	0
Food and supplies	550	650	100	2,000	2,500	500
Repairs/maintenance/cleaning	497	712	215	2,200	4,200	2,000
Registration	50	50	0	200	200	0
Equipment and stationery	50	150	100	550	550	0
Post/telephone/fax/mobiles	80	90	10	337	345	8
Total	11,427	14,852	3,425	47,687	57,195	9,508
Contribution	6,573	5,148	1,425	29,313	21,805	7,508
General overhead	4,600	4,500	(100)	19,725	18,000	(1,725)
Surplus/deficit	1,973	648	1,325	9,588	3,805	5,783

What this budget describes

In the example of the budget statement in Figure 6 you will see that it is subdivided into rows. These rows are known as budget lines and each row represents different items (staff costs, transport, equipment and so on). When the annual budget is agreed, resources are allocated to budget lines. The annual budget is then divided into monthly targets.

Actual expenditure is then coded to the appropriate budget line so that actuals can be compared against the budget, and the emerging position for the financial year can be seen. The budget statement sometimes shows the position projected to 12 months so that you can anticipate the end of year position if you continue to spend at the present rate. Please note that the statement example here has been simplified with the different items for expenditure combined for purposes of space and that your actual budget statement would have more lines for income and expenditure items.

- The budget statement is for a residential service and describes the income and expenditure for the month of July. The income and expenditure budget shown above has a series of rows and columns. The rows show the income (in this example, fees for the residential service) and expenditure (mainly staffing in this example).
- The difference between the 'actual month' and the 'budget month' is the 'variance'. The 'variance' column provides information on the under or over expenditure for the month.
- 'Actual YTD' (year to date) and 'budget YTD' figures describe the income or expenditure from the beginning of the financial year month of the budget to the month of the budget statement. 'Year to date variance' represents the accumulated difference between expenditure to date and the original budget plan to date and provides under and over expenditure for the year thus far.

Accruals

You may find that the figures in your budget statement do not match exactly with what you believe has been spent. This is because your organisation may prepare 'accruals', which account for items of expenditure that you have incurred but have not yet been paid for. An example would be quarterly gas and electricity bills, where you are using the supply each month, but only pay quarterly in arrears.

Variances

As previously explained, variances show the monthly and cumulative differences between the actual income/expenditure and budgeted income/expenditure for your service. There are a number of different possible reasons for the variances that may appear on your monthly budget statement.

1. **Staffing** – where costs have not been met because a post has not been filled or where you needed a greater level of staffing as service users' needs changed. The degree of control you have over these decisions will depend on your organisation and how much responsibility it delegates to the front line for recruitment. It is likely that you will have control over decisions about the use of agency staff or overtime.

2. **Volume** variances where the number of service users has increased or decreased over a period. It may not be possible to control this factor if demands on the service are higher than anticipated.

3. **Price** variances where the cost of supplies is higher than expected. A way you may be able to control this is by trying to achieve better terms from suppliers, although this will depend on your organisation's policies on dealing with suppliers (for example, you may have a central procurement process).

4. **Efficiency** variances are related to the level of resources used. For example, a residential home may need agency cover due to staff sickness. You may have little control over this situation occurring if it is caused by unforeseen circumstances, but you have to be able to influence its impact by choice of response, for example, when to cover through overtime, bank staff or agency staff.

Your organisation should inform you about strategies for managing your budget and reducing discrepancies and planning for the future. If you have any doubts or questions about variances you should contact your finance department. Finance managers value frontline managers who make regular contact and seek advice and guidance as this ensures that the budget is being managed efficiently and that the frontline manager feels accountable for their budget responsibilities.

Virement

Virement is the term used to describe an agreed transfer of money from one budget head – either income or expenditure – to another budget head. It provides the frontline manager with flexibility where resources that were allocated for a planned expenditure for a particular purpose are no longer required. It enables you to make adjustments in your budget that could not have been foreseen at the start of the year. You will have delegated authority to vire between budget heads.

If your budget shows a mix of overspend and underspend then you should have the possibility of balancing expenditure between different budget heads. Where you have a surplus under one budget line you may be permitted to overspend on another line so that the totals remain in balance. The rules regarding virement may vary in different organisations and you should not assume that you can vire between different lines of your budget without consulting your finance manager. Some income and expenditure cannot be transferred (vired) between budget heads as the funds are 'designated expenditure' and an underspend in one area cannot be used to compensate for an overspend on another.

Power to make budgetary decisions

Your main power will be in relation to revenue expenditure, particularly staffing, where you will need to make decisions within your budget responsibility about the staffing level of your service. You are also likely to have some limited power over minor capital expenditure – often for relatively small items, such as furniture and equipment – which will include the purchase of goods and services from outside the organisation.

Social care activities are labour intensive with a large percentage of total costs being wages and salaries. Staff will be on nationally agreed pay scales in local authorities and locally agreed pay scales in independent sector organisations, which reduces your flexibility to control your budget. Your freedom in budgetary terms will mainly be through the limited balancing of overspends and underspends between certain budget heads.

Getting support for managing your budget

It is important that you accept the accountability that accompanies budget responsibility. You should actively work with your finance manager so that you can control your budget and demonstrate a commitment to good financial management for your service. As a result of the greater delegation of budgetary responsibility to the front line of services you need to develop skills and knowledge about financial management that will enable you to participate in the planning and management of your budget (Martin & Henderson, 2001).

As a manager responsible for your budget you should feel you are trusted to manage the resources allocated to you. This can be difficult for finance managers as they have to strike a balance between overall responsibility for a service's financial health and obtrusive policing of individual budgets. Increasing the level of authority that managers can exercise, so that they have the discretion to spend resources, and helping them to understand that the budget is a *plan* based on assumptions that may change over time and are not a reflection on the manager if these assumptions turn out to be inaccurate (Palmer *et al*, 2002).

It is also important to remember that devolved responsibility for your budget does not mean you are on your own. It is helpful to view managing your budget as a collaborative activity rather than something you should struggle with on your own. Staff in your finance department are there to provide support and regular information, which means that you will need to ask elementary questions less frequently as time goes on. Their job is to support you and provide advice and guidance on the management of your budget. They can also provide training in the management of your budget and create a relationship where you can feel comfortable questioning financial support staff on any aspect of budget management. It is important to use these services, particularly if you have not managed a budget before.

Conclusions

Frontline managers are increasingly expected to take responsibility for the budget for their team or department. Ideally, they need to be involved in the budget-planning process and thereby contribute to decisions about the allocation of financial resources for their service. Becoming responsible for a budget means becoming familiar with the language of finance as this will help with interpreting monthly financial statements. Frontline managers are not expected to have the skills of an accountant, and should draw on the expertise of the finance team to support them.

References

Field R (1999) Budgeting is simple, it's management that is difficult. *Management Issues in Social Care* **6** (1) 1–6.

Gallop L (2003) Managing budgets and giving best value. In: J Seden & J Reynolds (Eds) *Managing Care in Practice*. London: Routledge.

Martin V & Henderson E (2001) *Managing in Health and Social Care*. London: Routledge.

Palmer P, Severn J & Williams S (2002) *The Good Financial Management Guide for the Voluntary Sector*. London: NCVO.

Managing at the Front Line: Managing information and communications technology

Key points

- **Information and communication technology is increasingly important in social care services.**
- **Frontline managers need to develop their skills in using computers and software as these can significantly improve the effectiveness of services.**
- **The use of the World Wide Web or Internet, has transformed the potential to access information on services development.**
- **The advent of information technology has also raised issues about security and ethical considerations about the use of information held on service users and staff.**
- **There are increased opportunities for managers and their staff to use computers for learning purposes with the development of online courses.**

Key words

Computers; information; communication; ethical issues; security; health considerations; email overload; e-learning.

The impact of information and communications technology (ICT) on social care services is significant and has become increasingly important, with the potential to manage information for purposes of planning, monitoring, and decision-making, and to communicate rapidly both within organisations and also externally. More recent developments with the use of laptops and hand-held computers gives managers and other staff who work in a range of locations the ability to communicate with their offices and colleagues.

Similarly, the rapid growth of the internet is providing managers with the opportunity to access policy documents and provide up-to-date information on new service developments. Similarly, access to research is now freely available online through dedicated social care websites. We have also seen local authority and independent sector organisations designing their own websites to provide information to staff, service users and carers.

In spite of these recent changes, the adoption of IT has been slow in social care with workers reported to be reluctant to embrace the new technology. The unskilled use of computers has meant that more time is taken up sitting in front of a computer than in face-to-face work with service users (Gregor, 2006). Some of the resistance to ICT can be attributed to suspicion of its relevance, with staff strongly committed to interpersonal relationships and who consider ICT in such a culture as a threat, although this is changing as it is recognised that properly used ICT is productive and effective in the delivery of social care services (Weiner & Petrella, 2007).

The importance of ICT skills is becoming increasingly urgent as social care services shift from traditional paper-based to computer-based systems. It is now a basic requirement for many jobs in social care that staff have ICT skills. These skills are needed for the purposes of inputting information on service users' records, writing reports, email communications and personal development through computer-based training and education programmes.

At a national level the General Social Care Council requires student social workers to be computer literate as a prerequisite for gaining a degree in social work (GSCC, 2002), and it is now becoming the norm for job descriptions to include applicants needing to have experience of using computer software related to service users, budgets and management information (Gregor, 2006). The development of the European Computer Driving Licence (ECDL) qualification has been validated by social care professional bodies as the main qualification and is increasingly seen as the basic qualification for staff in social care services. It is offered as part of in-house training by some local authorities, although there are regional variations in this provision.

As a frontline manager you need to develop an understanding of the uses of ICT and the skills necessary to use it effectively. These skills are becoming necessary as social care services have become more complex, with greater financial accountability, increased monitoring and regulation, which all require skills and understanding in terms of managing information.

Management information systems

One of the main developments of computers in social care, particularly in local authorities and larger independent sector organisations, is the use of management information systems (MIS) to enable the collection of data on a range of indicators, linked to the performance of services. For local authorities some of this information will in turn be used to respond to requests for data from central government. This information then feeds into public statements about the overall performance of services and whether they are performing well. For all managers high-quality information is essential as it can give an overview of a service, and how current performance compares with the past (Coulshed & Mullender, 2006). The key skill is managing this information so that it can be related to other sources of information (for example, observation, verbal reports and feedback from service users) to provide a fuller picture for the purposes of decision-making.

One of your roles as a frontline manager is to support staff who are required to collect and record information as this can play a crucial role in informing a service about users' needs, patterns of delivery and what works and what does not. You have an important role in supporting staff who may not be motivated to collect and record information if they cannot see the benefits to be gained for their efforts, particularly when it competes with other practice demands (Ousley *et al*, 2003).

Information can broaden a team's vision of their work and the issues they need to grapple with. Without current and relevant information and actions, decisions are likely to be based as much on assumptions and prejudice as on knowledge (Hearn *et al*, 1992). In information terms, you are the bridge between practice and management, which means you are located at the nexus of one of the most important communication points in the service. You will also be required to provide information for senior managers and it is important that you evaluate the information collected and assess its usefulness, ensuring that it is commensurate with the effort required to gather it. It is important to see data collected as a helpful way of managing practice rather than simply a tool of management (NISW, 1996) and something that can be used to meet the needs of managers, staff and service users (Coulshed & Mullender, 2006).

One of the problems in using management information systems is that different organisations have their own computer software and their systems may be incompatible. This can be a major obstacle, for example, in single assessments between health and social care services, where different systems reduce inter-agency effectiveness. The importance of shared electronic recording processes is important for the quality of joint working (Torkington *et al*, 2004).

Social care services websites

One of the rapid areas of development that has been facilitated by the growth of websites is the information available to the public about social care services. In some areas these websites are well developed, whereas others are limited and the websites difficult to use. If the NHS experience is a representative example it is clear that service users and carers will increasingly use websites to seek information. This will mean that the relationship between services and their users will change as the power rooted in information is no longer held solely by professionals.

Internet or World Wide Web

The Internet is a huge network of computers that spans the globe and is a system that allows the exchange of information between computers. Because of the range of services available it is increasingly valuable as an information resource and communications superhighway. For example, you are able to access thousands of databases at government departments, universities and research centres throughout the world, read electronic journals and exchange information, and send virtually instantaneous electronic mail (email) to any organisation or person connected to the Internet. The Internet offers you the potential to considerably expand your knowledge and understanding through access to websites that are designed specifically for social work and social care professionals, and associated sites, such as those providing information on the NHS.

The Internet is seen as an untapped information resource in social care with the potential for improving work skills through access to e-learning and websites. Although over 80% of staff surveyed by MORI for SCIE (2007) in research on the potential for e-learning, had access to a computer at home, the use of the Internet was only mentioned by 28% of staff and was only regarded as the best source of information on working better by 18%.

An enormous range of information is currently available, with the main access points for social care practitioners and managers being through websites such as those listed below.

- Department of Health provides a wealth of information on all aspects of social care, including, national policy, practical guidance and service standards – **www.dh.gov.uk**
- General Social Care Council (GSCC) is the social care workforce regulator for England and provides information on the Social Care Register, codes of practice and training and learning – **www.gscc.org.uk**
- Social Care Institute of Excellence (SCIE) has its mission to identify and spread knowledge about good practice in social care – **www.scie.org.uk**
- Skills for Care is responsible for the development of standards and qualifications for staff working in adult social care. It consults with employers, service users and carers to modernise social care in the UK – **www.skillsforcare.org.uk**
- Social Care Online is the UK's most complete range of information on all aspects of social care research and is free to users – **www.socialcareonline.org.uk**

> *'Since I developed my skills in using the internet I've started accessing the Department of Health website and have been better informed about government policy changes and have also been able to keep staff informed. I'm no longer dependent on my department feeding information to me. I've often got the information before my manager now!'*
>
> Frontline manager, group care services

Electronic mail (email)

Alongside the World Wide Web, the use of email is one of the main attractions of the Internet. It allows for information exchange in the form of messages and, where necessary, attached correspondence, files, reports and graphics to be sent worldwide in seconds.
It is a fast and effective way of communicating within a service and externally and has largely superseded paper-based systems in many organisations. Its greatest value is the ability to send information and reports accurately and quickly with replies often received within minutes.

Email is now widespread and is the same as a letter or written message sent by post, but exchanged in a different way. It is increasingly used as the main means of communication in social care services, with many benefits such as the ability to send messages at any time of the day or night from any location. It is also used to send documents as attachments which is increasingly replacing post and fax machines as a means of document distribution.

The downside of email is communication overload where managers are swamped by the number of emails in their 'inbox', which can consume considerable time reading and responding to this mail. Emails also feel more immediate and urgent to the recipient. Much of the problem is the tendency for senders to include everyone in their distribution and to demand instant responses, unknown in the days of telephone, post or fax. The management of email is important if this valuable tool is not to be misused.

Because of the problems of overload experienced by managers the following guidance has been developed as good practice for email use (Kakabadse *et al*, 2004).

Do:

- make sure that the title in your message tells recipients what it's about
- use markers to show importance (high, normal, low)
- use indicators and dates in the subject line (for information; for action)
- show consideration for others' working patterns
- keep attachments to a minimum.

Don't:

- use email when phone or paper might be better
- automatically reply to all recipients – limit copies to those who really need to know
- send emails late at night expecting a reply the first thing in the morning.

It is also important to remember that emails are disclosable under information access legislation and fall within the scope of the Data Protection Act (1998), so it should be used with caution when discussing employer performance or other sensitive information.

Intranets and extranets

Many organisations are developing 'intranets' which use the same technology and protocols as the internet, but are restricted to specific users within a particular organisation or other approved organisations which own the intranet. These systems are well developed in local authorities and the NHS where staff can communicate rapidly on a range of information to speed up decision-making. This type of information exchange using intranets is increasingly a feature in all public services. It has the benefits of providing staff with a common knowledge base and can improve flexibility and overheads by allowing staff to work from any location. The extranet is similar but is made accessible to selected organisations and staff for the purposes of exchanging data and sharing information.

Microsoft Office

The Microsoft Office suite of programmes is widely used in social care services and has a wide range of management applications. It is particularly important to develop your skills in the use of these programmes as they are routinely used by managers in organisations and are a required skill for some management posts. The suite contains the five programmes:

- **Word:** used in word processing for reports and letters and is the standard program for these applications
- **Excel:** used for numerical analyses such as creating and manipulating financial spreadsheets and budget reports – your organisation's budget reports will be put together using Excel or a similar program and you need to become familiar with its use
- **PowerPoint:** used to create computer presentations, such as slide presentations – skills in using this program are essential as you will be required to make high-quality presentations to a diverse range of audiences
- **Access:** used for database management purposes for collecting and manipulating data and is important in the preparation and monitoring of management information
- **Outlook:** used to manage personal information, addresses and phone numbers and email.

Security and ethical considerations

The use of computers raises questions about the security of information, how privacy can be protected and who has access to data. You must ensure that all your information systems comply with the Data Protection Act (1998). This act ensures that organisations manage the personal information they hold in a responsible way and that confidentiality is respected and not misused. This act also requires that all service users are aware that information being held on them is held electronically. You should also be aware that the Freedom of Information Act (2000) allows service users to view the records held on them by a public service, with certain exceptions. This should caution staff to be particularly vigilant when recording information that may turn out to be inaccurate, but which could be accessed by service users if they used their rights to see information held on them. This is all the more important as Sapey (1997) has stated that one in four records held by social services departments contained inaccurate information. This means that managers need to ensure that staff pay attention to the accuracy, completeness and relevance of records, just as they would for paper records. This risk is only likely to increase as workers directly input information electronically (Ousley *et al,* 2003).

How information is accessed, recorded and stored needs to be part of your organisational policies with staff trained in the practices that conform to the legal requirements. Data held on computers can be accessed only via electronic passwords for security purposes. You need to ensure that you have a policy that informs your staff how they record and store information as part of their contractual code of practice and that it conforms to the act.

A further security risk is the potential for virus infection of computer systems. Viruses are intended to harm computer systems and they can be spread in many ways, typically through attachments to emails and when downloading files from websites. It is essential that your service has an up-to-date anti-virus programme and a firewall to stop viruses getting into your computer system.

Health and safety

When personal computers were first introduced little thought was often given to the consequences of lengthy computer use on staff. Concerns over eye strain, wrist strain and back-strain have become more prevalent and have led to the development of advice on healthy work environments. See **www.flexibility.co.uk** for tips on using computers safely.

It is now clear that the environment where computers are used is important and that their use for concentrated periods without breaks is unhealthy. Similarly adaptations to computers to make them more comfortable to use include the use of screen filters and specially adapted keyboards. On-screen changes can include the use of different sizes of icons and fonts, particularly for staff with visual impairments. See Gregor (2006) for a checklist regarding the safe use of computers.

The Management of Health and Safety at Work Regulations (1999) require employers to provide training for staff when computers are introduced. The Health and Safety (Display Screen) Regulations (1999) also requires all computer users to be risk assessed related to their use of computers, with the provision of equipment where necessary. The Disability Discrimination Acts (1995) and (2005) require employers to make adjustments to support people with disabilities so that they can carry out their jobs effectively, including providing adaptations to computer equipment where necessary.

E-learning

There is considerable potential for improving the skills of staff through the use of computer-based training (CBT) programmes. These programmes are seen as flexible in terms of time and can be cost-effective with staff having the opportunity to work at their own speed as a key benefit (SCIE, 2007). Staff using e-learning programmes demonstrated that it could have a number of benefits including:

- *'work at a time, place and pace that suited them'*
- *'facilitated collaborative learning through chat rooms and peer-group support'*
- *'built confidence through reducing fear of making public errors'*
- *'helped build links between isolated individuals'*
- *'provided access to expertise and resources irrespective of distance'.*

An example is the growth in 'paperless NVQs', using e-portfolio software, which provides greater flexibility with staff able to work on their evidence at times convenient to them and from locations of their choice. Staff input information into their computer using password-protected access sites and much of the communication with the assessor is via email.

One of the difficulties identified in the SCIE research (2007) was that social care is now delivered by a large number of small private organisations where learning was not embedded in the organisational culture, and consequently they are not likely to be ready to supply or promote e-learning, in spite of staff willingness to engage with this form of learning. E-learning was found to be limited in the social care sector in spite of staff having a high degree of IT literacy and who may actually be more ready for e-learning than employers think (SCIE, 2007).

Developing your skills in information technology

Traditionally, staff training in social care services were seen to be resistant to developing computer skills, although these have now become a necessity as so much of the work of social care organisations is based on information and communication contained on computers. Some of the resistance was said to be because computers offered little to staff and that they were people orientated and did not see computers as adding benefit to their work.

Some of the resistance to the use of computers has been overcome with the introduction of computer training on social work degrees and in-service training courses, particularly in larger organisations for staff who are completing work-based social care awards. The development of the European Computer Driving Licence (ECDL) qualification is now recognised as providing staff completing this award with competence in computer skills. Training for the ECDL is also provided by some local authorities, although there are variations in the availability of this training in-house. Information on computer training can be found on the websites of local further education colleges.

You also need to develop your skills in computing as this is a skill that will become increasingly important for communications and accessing information that can help you to make better quality decisions and enable you to transmit and receive information much more quickly than paper-based systems. A further important skill is the use of the Internet. This will enable you to access information that can help you in your decision-making.

Therefore, in terms of current operational requirements and in terms of longer-term career development, it is in your interest that you become confident in the use of computers and that you see them as another tool that can assist you in effective management. Much of what you can achieve in this area will be dependent on the information strategy developed by your organisation and on how far it has progressed in training and supporting managers and other staff to develop skills in the use of computers.

Irrespective of the rate of progress in your department (and many managers told me of their frustration at the slow adoption of information technology to enable them to perform more effectively) there is much to be gained by taking the initiative and developing your skills in this area. This will enable you to comment critically on information management strategies from an informed perspective and to support new initiatives to enhance information systems.

Conclusions

Information and communication technology is increasingly impacting on social care services through the use of management information systems, the Internet and email. The growth of this technology has in turn raised issues about the security of information, and also ethical and health concerns that need to be considered. There is also risk of overload as a result of the almost universal use of email for communication purposes. The development of skills in using computers and their software programmes is essential for frontline managers and their staff. Computer-based training programmes provide a convenient resource that should be used to supplement other forms of training.

References

Coulshed & Mullender (2006) *Management in Social Work*. Basingstoke: Palgrave Macmillan.

General Social Care Council (2002) *Accreditation of Universities to Grant Social Work Degrees*. London: GSCC.

Gregor C (2006) *Practical Computer Skills for Social Work*. Exeter: Learning Matters.

Hearn B, Darvill G & Morris B (1992) *On Becoming a Manager in Social Work*. Harlow: Longmans.

Kakabadse A, Bank J & Vinnicombe S (2004) *Working in Organisations* (second edition). Aldershot: Gower.

National Institute for Social Work (1996) *The Social Services Information Agenda, Briefing 17*. London: NISW.

Ousley M, Rowlands J & Seden J (2003) Managing information and using new technologies. In: J Seden & J Reynolds (Eds) *Managing Care in Practice*. London: Routledge.

Sapey B (1997) Social work tomorrow: towards a critical understanding of computers in social work. *British Journal of Social Work* **27** (6) 803–814.

Social Care Institute of Excellence (2007) *E-Readiness in the Social Care Sector: Building the capacity for e-learning*. London: SCIE. Available from: www.scie.org.uk.

Torkington C, Lymberry M, Millward A, Murfin M & Richell B (2004) The impact of shared practice learning on the quality of assessment carried out by social work and district nurse students. *Learning in Health and Social Care* **3** (1) 26–36.

Weiner ME & Petrella P (2007) The impact of technology: implications for social work and social care managers. In: J Aldgate, L Healy, B Malcolm, B Pine, W Rose & J Seden (Eds) *Enhancing Social Work Management*. London: Jessica Kingsley Publishers.

Section 4

Developing yourself, your staff and your service

Section 4

Chapter 18

Managing at the Front Line: Managing your own development

Key points

- **Management at the front line places heavy demands on the individual and developmental opportunities are important in supporting the manager.**
- **Management development should be a central part of the work of the frontline manager and integral to their day-to-day work with its learning potential recognised.**
- **Organisations should ideally integrate management development and the strategic aims of the service so that managers have the skills to help implement policies effectively.**
- **There is a wide range of development activities that frontline managers can engage with and these can be tailored to suit the individual's needs.**

Key words

Messy problems; commitment to development; reflecting on practice; using a range of development opportunities.

We have seen in Chapter 1 that frontline managers are considered the 'keystone' in a social care service and that their quality and competence make a significant difference to an organisation's performance with strong professional leadership and management essential to implementing social care objectives (Skills for Care, 2006). There is growing realisation that there is a close relationship between the quality of those managing services and practice that is valued by people using services (Statham, 2004). Because of the demands made on frontline managers it is essential to strengthen and develop their skills and knowledge, ideally within a learning culture that provides a range of opportunities for the manager.

Management as a unique activity

Much management has been described as an activity that is primarily concerned with uncertainty in organisations, and this has important implications for what managers need to learn. Although there are some activities that are predictable, despite what is described in management textbooks, most management activity is less structured, planned or co-ordinated than ideal type models would have us believe. Management is essentially about dealing with the messy problems that are left when those tasks that can be dealt with by routine, specialist, technical and professional work are removed (Burgoyne, 1999). It is this messiness which makes management different from the point of view of learning.

The learning and development needed in order to prepare managers for less structured work is different from the learning applied to routine problems. They can be addressed primarily through technical training to deal with specific activities. The management role calls for a range of human abilities, including creativity, judgement, intelligence and insight. These abilities are the generally less specific and are 'developed' rather than trained. This view of management means that approaches to development need to encompass a wide range of opportunities. These should include formal activities such as training programmes, but also mentoring and coaching, and access to learning resources such as e-learning materials. At its heart, management development should be an integral part of management and concerned with how you can improve your effectiveness and the performance of the organisation using a wide range of learning processes applied to all aspects of your work.

Organisational commitment to management development

How your organisation responds to your learning and development needs will be dependent to a great extent on its strategy for developing its staff and the integration of staff development with the corporate goals of the organisation. The greater the degree of integration between the development of strategy and management development, the more likely it is that managers will be able to meet the challenge of implementing the organisation's strategy.

Development needs at different career stages

The CIPD (2008) have identified three broad situations that occur around the move into management roles that require different development opportunities.

1. When frontline managers are first appointed they need *preparational development* activities that offer them access to the new skills they need when promoted to a management role. These ideally range from specific training courses on technical aspects of the role to the wider management skills needed.

2. There is *transitional development* where a frontline manager needs specific support, such as coaching or mentoring, as they adjust to the new role and the transition, including the more complex issues of personal change arising from dealing with the new role and responsibilities.

3. There is *role development* where the frontline manager who is new to the role may need help in adjusting to the new demands and may need support in making a satisfactory transition from their old role and begin to adjust to the new demands and the learning this demands.

Some management development techniques

In discussing the range of development activities potentially available to the frontline manager, these can be divided broadly into those that are work-based and those that are based on a formal training and education programme, although this is not a clear cut distinction as attendance at a training course should also provide opportunities to apply the new skills learnt in the workplace. Where the emphasis is on people skills and interpersonal behaviour, training courses are less effective and work-based methods are more effective.

Informal learning

Managers learn and improve their skills all the time from their experience of doing their jobs. A great deal of management development occurs in this way. For example, a manager has an unplanned experience, this leads to discussion between two managers, which reviews the action, and a decision is made about what to do in a similar situation in the future. These are all potentially elements of an effective learning cycle, although managers will rarely recognise at the time that they are learning – they think they are simply 'managing' (Mumford, 1996). It is when the experiences of the job are acknowledged and used to inform future actions that it becomes a management development activity.

Training courses

These are formal training courses that address skills deficits related to technical aspects of a frontline manager's role, such as budget management through to delegation and performance management. These courses are more usually delivered through short programmes or modules with work-based activities between sessions in order to increase the potential for the effective transfer of course-based instruction to the workplace. The shift to a shorter modular structure has grown out of the recognition that managers learn best from short intensive programmes that are interspersed with practical experience. A crucial test of the effectiveness of training is the potential for transfer of skills learnt in the training environment to the workplace and how these new skills can be retained with support and opportunity to use them. Key support activities could include coaching by other managers, goal setting, reinforcement activities and the modelling of behaviours (Bratton & Gold, 2003).

Coaching and mentoring

These are both one-to-one methods that offer personally tailored reflection and discussion in confidence between a frontline manager and another individual focusing on the manager's development. Coaching will focus on improving a manager's skills and performance, with the coach, often from outside the organisation or a more senior manager, using a structured dialogue to identify potential solutions to problems or issues that the frontline manager is dealing with.

Mentoring is usually provided within an organisation and is typically where an experienced manager (who is not the individual's line manager) meets regularly with the frontline manager to help them to perform better in the role. It differs from coaching in its focus on longer-term learning and development, and is less focused on events than coaching.

Action learning

Action learning was developed by Revans (1982) in which a group of managers (an action learning set) meet regularly over time to discuss the individual members' real-life management issues with the aim of learning from each other. To work effectively the set needs to be skilfully facilitated so that members focus on solving the issues occurring in their workplace, by trying out different approaches, with discussion and support from colleagues to help them reflect on their impact. Revans originally saw it as a means of dealing with novel problems that did not have easy answers and emphasised the importance of asking good questions in order to challenge current ways of dealing with management problems. It is seen as a particularly helpful form of development for managers who can be isolated in their normal day-to-day work and who benefit from group involvement with other managers.

Development reviews

Many managers will be familiar with appraisals that have become a regular feature of organisational life, annually recording performance and developmental needs in a top-down process. Developmental reviews differ in that there is an open and two-way dialogue which focuses on the manager reflecting on their performance as the basis for making a development plan. The difference between the traditional appraisal and the development review is that the former concentrates on performance, whereas the developmental review has the development of the manager as its main focus.

360 degree performance review

The growth of 360 degree performance reviews is intended to overcome one of the main problems with any appraisal or performance review, which is that it can be perceived as subjective and over-dependent on the views of the person conducting the review. The administration and interpretation of a 360 degree review needs a skilled coach or consultant experienced in the use of management development. It is based on the assumption that self-awareness increases when feedback is based on a number of assessments and that it is a prerequisite for improved performance. This form of review typically incorporates voluntary feedback from managers, peers and members of staff. It is usually confidential and learning-oriented rather than linked to assessment and is most effective when it is combined with coaching on developing self-awareness and behaviour management. It is claimed to have other benefits such as involving staff in upward appraisals and is said to enhance commitment to an organisation and fits with the notion of employee involvement and open organisational cultures (Bratton & Gold, 2003).

E-learning

E-learning refers to the wide range of material available on websites and through email with computers used to deliver training. Some of the material is open access and some is now delivered through organisational intranets. This approach has advantages such as flexible access where managers can learn in their own time. It is considered to be more effective for 'hard' knowledge than for softer skills such as communication or leadership. It can also have benefits for managers who prefer the anonymity of the internet where they can explore the issues on their own without the direct involvement of others. Due to the plethora of material on the internet, skilled guidance is required to make use of this form of learning.

Shadowing

This is a very effective way for new managers to learn from a colleague through observation and feedback. Typically a manager would spend a day with a colleague in a similar role and at the end of the day there is a de-briefing where the 'shadower' feeds back to their colleague about what they have observed. The roles can then be reversed for a day. It is a highly cost effective way of learning, but needs thought and preparation if the time spent is to be productive in terms of activities that can offer learning opportunities.

Formal management qualifications

In addition to the work-based activities that constitute the main elements of management development there are also a range of formal qualifications that are available to managers. If you are a manager in adult services it is now a requirement that you complete the Registered Manager Award (RMA), which is a competency-based qualification that meets the standards for best practice in social care. The RMA is offered by a wide range of training and education providers using various forms of assessment that are suited to the needs of individual managers.

Managers who already hold professional qualifications in management such as the RMA may wish to progress their career and complete an undergraduate or postgraduate qualification in management offered by a range of providers such as colleges, universities and private training providers. Some of these qualifications will be part of the post-qualifying (PQ) framework for social care staff and offer graduates and those with other recognised qualifications the opportunity to complete postgraduate awards ranging from certificates through to masters degrees (see the General Social Care Council website containing details about the different PQ awards, including those that focus on leadership and management).

Continuing professional development (CPD)

In a world of rapidly changing social care practice it is essential that skills, knowledge and experience are regularly updated. It is particularly important as managers can be reluctant to engage with learning activities and organisational attitudes towards learning and development can reinforce such attitudes (Preston-Shoot, 2007). Continuing professional development (CPD) is seen as essential for the maintenance of a skilled and knowledgeable workforce. Effective CPD takes place within a learning culture with the structures and processes that enable the social care workforce to develop in their roles and keep their practice up to date. A well-developed CPD strategy is seen to support better recruitment and retention of staff and, in turn, better services. All registered social workers and managers are required to keep their training and learning up to date in order to re-register with the GSCC after their initial three-year period of registration. CPD ideally integrates learning with practice and is largely self-directed and relevant to the manager's own developmental needs, and can include conferences, training courses and completing PQ awards (Tovey, 2007).

Identifying learning opportunities

We have seen that there is a wide range of developmental opportunities open to you in your frontline manager role. Not all of these activities and opportunities will necessarily be available to you in your organisation, particularly if you work for a small service that does not have the infrastructure to provide the management development opportunities available in larger organisations. Nevertheless, it is important that you are aware of the range and potential of management developmental activities and that you are in a position to argue for the inclusion of these activities in the human resource policy of your organisation.

Explore the different approaches potentially available to you with your manager, and where possible, with an experienced management development adviser. This will allow you to identify those learning opportunities that best fit your preferred ways of learning and personal circumstances, and to establish what is practical in terms of access and participation in development activities both within the work setting and externally through educational and training programmes. Managers interviewed for this handbook suggested that the ideal development activities combined both vocational and academic elements, providing space to reflect on the complexity of management in a social care setting and also offering practical ideas that can be tested out in the workplace.

Frontline managers who responded to the survey identified a wide range of activities that had been helpful in their development, including:

- *'watching other managers and how they handle situations'*
- *'learning from my manager'*
- *'learning from role models'*
- *'making mistakes and learning from them'*
- *'being a reflector'*
- *'attending training courses'*
- *'learning how far to push myself when faced with new challenges'*
- *'using supervision to look at my performance'.*

Conclusions

Management development is essential if frontline managers are to be able to continue to respond to the demands made on them. As services change and develop, frontline managers similarly need to identify their learning needs and seek out activities that can help them respond to change. Where management development is integrated with the strategic aims of an organisation, frontline managers will be better prepared to manage change.

References

Bratton J & Gold J (2003) *Human Resource Management, Theory and Practice* (third edition). Basingstoke: Palgrave.

Burgoyne J (1999) *Developing Yourself, Your Career and your Organization*. London: Lemos and Crane.

CIPD (2008) *Management Development Factsheet* [online]. London: CIPD. Available from: www.cipd. co.uk (accessed January 2009).

Mumford A (1996) How managers can become developers. In: J Billsberry. *The Effective Manager: Perspectives and illustrations*. London: Sage.

Preston-Shoot M (2007) Engaging with continuing professional development: with or without qualification? In: W Tovey (Ed) *The Post Qualifying Handbook for Social Workers.* London: Jessica Kingsley Publishers.

Revans R (1982) *The Origins and Growth of Action Learning*. Bromley: Chartwell-Brat.

Skills for Care (2006) *What Leaders and Managers Do in Social Care: A statement of leadership and management development strategy for social care*. Leeds: Skills for Care.

Statham D (Ed) (2004) *Managing Front Line Practice in Social Care*. London: Jessica Kingsley Publishers.

Tovey W (Ed) (2007) *The Post-Qualifying Handbook for Social Workers*. London: Jessica Kingsley Publishers.

Managing at the Front Line: Managing staff development

Key points

- **Staff need to continuously develop their knowledge and skills to ensure that their practice keeps pace with changes in social care services.**
- **Frontline managers play an important role in supporting, facilitating and identifying learning opportunities for their staff.**
- **The provision of work-based training is one of the most effective means of enhancing the practice skills of staff.**
- **Frontline managers' own attitudes to learning can set an example to staff and contribute towards building a learning culture in a service.**

Key words

Knowledge and skills; learning opportunities; influence of frontline manager; different learning activities.

Frontline managers have a critical role to play in staff development. They should focus on the learning needs of their staff, with the aim of enhancing their performance to meet the needs of service users. Increasingly, staff development is moving away from the focus on individual training activities to one that is built into the fabric of how staff are managed and is one of the key ways of meeting organisational objectives (Coulshed & Mullender, 2006). Frontline managers are at the centre of this development as they are responsible for organising and supporting their team's learning opportunities.

Why is staff development important?

Evidence from research and evaluation studies have identified the critical roles that knowledge and skills play in enabling staff to work effectively in rapidly changing social care services. One study argued that the social care workforce is unable to re-skill at a rate commensurate with the pace of change in services, particularly in relation to the work with service users who have complex needs (Balloch *et al*, 1999). A further study identified that some social care staff were out of date, unclear how to update their knowledge and skills and were offered limited training opportunities (Braye *et al*, 2006).

At a wider level, organisations have to continually adapt to changes in their environment in order to avoid deteriorating performance. To achieve this they depend on their staffs' ability and willingness to learn and, in turn, change their work practices. This recognises that staff have to be prepared to change as a result of learning and development opportunities so that their service can keep pace with changes in the external environment. Organisations need to work on the assumption that if staff are not developing as fast as changes in their environment then they are losing the capacity to be effective (Hudson, 2002).

Although many staff in social services departments will have undertaken formal educational and training programmes to achieve professional qualifications, it is in the support of work-based learning that frontline managers have a critical influence. The findings of the Learning Society Programme states that of all the mechanisms used in organisations to promote learning, the most important is the appointment and development of managers and their influence on the service and their approaches to learning (Eraut, 1998)

This research also suggested that formal education and training provide only a small part of what is learned at work. Most learning was described as informal, neither clearly specified or planned, and arising naturally out of the demands and challenges of the work (eg. solving problems, improving quality or productivity or coping with change) and out of social interactions with colleagues or those using a service. This work-based learning could nevertheless be enhanced when it was experienced in a supportive climate, created by the manager, to facilitate this informal learning.

Frontline managers are influential in a number of ways. Smale (1998) sees them as crucial opinion leaders who play a key role in any changes in practice or policy initiated by senior managers, or by frontline staff in partnership with service users. Similarly, their professional behaviour provides a model for staff, whose practice will be influenced by the standards set by their manager and their expectations of others.

Alongside these developments, evidence from a survey in 2004 conducted by the Chartered Institute of Personnel and Development (CIPD) has identified on-the-job training as one of the most effective ways in which staff learn in an organisation, followed by coaching and mentoring. Although formal training courses contribute to the effectiveness of staff, overall work-based activities, alongside experience of the work itself and informal support from colleagues, constituted over 80%, with only the remaining 20% derived from formal training courses.

Learning and development is now recognised as something that needs to be increasingly integrated with workplace activities, particularly as off the job training can suffer from the problem of transfer when what is learned externally is difficult to apply in the workplace, without detailed planning. Where the frontline manager creates opportunities for critical reflection with staff following training events this can enhance the value of off-site activities.

The frontline manager's role

It is important that frontline managers create the conditions in their teams where professional development is valued and seen as an integral part of the work of a service. Although increasing numbers of staff in social care services will have undertaken formal educational and training programmes to achieve professional qualifications in social work, for many social care staff the provision of learning opportunities in the workplace will be crucial to their development. The frontline manager sits at the centre of a system that has to balance the needs of individual team members, with the needs of the wider team and the overall requirements of the organisation (Peel, 2003). The frontline manager has to understand their responsibilities as all these elements are interacting and impacting on each other.

Frontline managers' involvement with staff development also has other benefits. Lower staff turnover, improved motivation and increased opportunities for promotion are seen as direct benefits. A further important reason for developing a strong workplace learning culture is the evidence that job satisfaction and staff retention is directly linked to training and development.

Alongside these factors is the shift from training to learning, where the focus is less on trainer-centred activities to one where a wide range of interventions and processes are used to create a climate that supports effective learning. This can be seen in the development of coaching, mentoring, learning sets and other work-based methods. All are intended to create a culture of the learning in organisations. Although learning is a discretionary activity and firmly in the control of the individual nevertheless frontline managers, supported by others, such as HR specialists in larger organisations, can create the conditions, structures and processes that empower staff to become committed learners.

The influence of culture on learning

Organisational culture has a powerful influence on how a service is managed and how staff respond to demands made on them, particularly related to change. In a CIPD survey (2004) questions were asked about the ways in which organisations could promote learning with culture and the commitment and skills of managers were overwhelmingly seen as the most important factors.

It is worth listing the factors that were seen to be the most important as these have lessons for social care organisations and frontline managers who are interested in developing their staff:

- ensuring that the organisational culture is supportive of learning and development
- ensuring that managers are committed to, and have the skills to support learning and development
- providing staff with time to learn at work
- ensuring that staff understand the direction and strategy of an organisation and how this relates to their development of skills
- providing a variety of learning options to suit staffs' personal learning styles

- providing advice to staff on how they can manage their learning
- understanding what motivates staff to learn
- rewarding staff who develop their skills and abilities
- having innovative and committed HR and training professionals.

Leadership in learning

As the frontline manager you possess considerable influence in promoting the vision of a learning culture in your service (Darvill, 1997). You also have the potential to influence a strategic approach to the development of your staff team to establish a clearer understanding of what the team is doing and how it could be improved for the benefit of service users. You also have an important role in setting an example as a learner. You can demonstrate this by:

- declaring the importance you place on your own learning
- identifying, with help where needed, your own learning needs
- asking difficult questions in supervision and team meetings that question both your effectiveness and the effectiveness of staff, and demonstrate a receptiveness to new ideas
- using formal and informal learning activities yourself.

Setting an example

You can also create a learning climate in your organisation by modelling leadership in learning through always engaging in learning something, sharing and demonstrating your new learning, for example, communicating with colleagues that you have learned something new, and making learning normal and legitimate, and encouraging others to do it. Darvill warns though that you should not feel under pressure to be competent all the time, and that you should avoid getting into a position of never admitting that you need to learn so that no one around you is willing to offer help.

Learning approaches

It can sometimes be difficult for frontline managers to think of creative ideas that can be used to develop their staff. One way is to use the following list to think of a problem that members of staff bring to you regularly and come up with a range of options that do not include sending the individual on an external course or conference attendance (which may of course be the correct choice for some subjects, when it is necessary to have expert tuition or information provided by informed speakers). Using a list of learning options can begin to open the eyes of staff members to the wide range of learning methods available.

Individual activities could include:

- reading
- writing
- keeping a learning log

- use of CD-roms and DVDs
- distance learning
- computer-based learning.

Informal, less structured activities could include:

- discussion with a critical friend
- observation of practice
- questioning a colleague
- consulting with a line manager or specialist.

More formal, more structured activities could include:

- coaching and mentoring newly appointed or junior members of staff
- discussion group of staff on specific topics of interest to the team
- learning groups or sets convened for developmental purposes
- discussion in a team meeting
- using supervision sessions for developmental purposes
- deputising or acting up to gain experience
- job rotation or secondment for a fixed period to widen experience
- shadowing a colleague with particular expertise
- visiting other services to examine different models of care or practices.

Possible approaches in work-based learning

In order to support work-based learning it is important to understand that there is a range of possibilities potentially available to staff. Again I have relied on Darvill's (1997) helpful review of learning methods, some of which will be familiar to you, and others which are more specialised and require additional support and expertise.

- **Induction** – training provided for new staff or staff transferring to new roles, with identification of learning needs that require further support.
- **Supervision** – seen as important in terms of supporting the learning needs of staff and can incorporate training techniques as part of the supervision process. The training focus of supervision is useful in helping staff to understand practice issues and in improving their skills.
- **Coaching** – showing staff directly how to carry out tasks, and providing close, reflective feedback on staff's work. This method is used to develop performance in relation to specific roles and tasks, through instruction, demonstration and practice. Frontline managers can coach, although there is also value in using staff who are not in a line management relationship where there is assessment of performance and developmental issues.
- **Mentoring** – this has some overlap with supervision and coaching, but it is less role or task-specific. It is essentially about providing support (particularly from someone who is experienced) and can provide a wider perspective – the 'wise elder'. It has been developed in many organisations in relation to the development of managers,

although frontline managers may increasingly be involved in mentoring staff who are studying open learning materials or completing post-qualifying awards that are based on practice evidence from the workplace. Mentoring, which focuses on the individual's development, is usually a more formal relationship than coaching. Mentoring can also involve someone who is not the person's line manager as it involves developmental rather than task performance.

- **Assessor roles** – these are well developed in social work with practice teaching as part of BA in Social Work courses and the assessment of national vocational qualifications in the workplace.
- **Team meetings** – these can provide the potential to be used as the basis for learning, particularly if they are facilitated by someone who does not have managerial authority for the team.
- **Lunchtime talks** – with either a staff member presenting an aspect of practice or invited speakers with particular expertise relevant to the team.
- **Inter-agency meetings** – the growing importance of partnership-working opportunities to bring together staff from different agencies (sometimes using open learning materials) has a history going back to the community care implementation of the mid 1990s. The mixing of staff groups from different professions and agencies is the most valued aspect of this manner of working.
- **Learning sets** – can be based on staff from one agency or a number of agencies meeting to explore current problems and producing results based on analysis of the problem and problem-solving ideas. They require close attention to the contract between the members and to ground rules, issues of time and the commitment of members. Skilled facilitation can also enable the set to achieve its aims more effectively.
- **Consultancy** – usually refers to an external consultant who works alongside managers and staff to support the introduction of new policies or other initiatives. The objectivity and breadth of experience that the consultant offers is an important factor in the decision to engage someone from outside the organisation. Frontline managers often act in a consultancy role in their own teams, although it is helpful in some situations if the consultant does not have line management responsibility.
- **Job rotation** offers the individual the opportunity to broaden their experience and skills in a time-limited attachment.

Recognising the pressures facing frontline managers

Frontline managers in social care services work under intense pressure as a result of the highly regulated and target-driven culture of public services, or in those independent sector provider organisations working on contract to public services. Modern organisational structures and processes have witnessed increasing amounts of responsibility devolved to the frontline of services, with many managers not in a position to play the traditional training role with their staff.

What managers can do is to act as a facilitator and enabler of training opportunities, rather than a direct trainer in the traditional sense themselves. What is important is the frontline manager acting as a 'gatekeeper' to a range of learning and development opportunities for staff. As it is the frontline manager who is likely to know individual members of staff well and have an informed view of their developmental needs, they are in a particularly strong position to facilitate development opportunities for their staff.

The other risk identified by Hutchinson and Purcell (2003) is that managers can be so focused on the short-term and immediate issues that it can drive out the important focus on the longer-term development of staff. This allied to a lack of time and financial resources for staff development can combine to act as a powerful barrier to staff development in a service. Learning and development can also be under valued when a team is surrounded by demands to deliver more services, manage change and resolve endless short-term crises, and as a result the need to invest in learning can be neglected (Hudson, 2003).

The importance of frontline managers having a clear understanding of their role in staff development, supported by other specialists, is essential if services are to maintain their capacity to meet the demands made on them. Work-based learning also needs middle and senior managers to make a commitment to support it and set an example through their own actions. Neglecting staff development can result in deteriorating levels of service, and impact on the morale and commitments of staff.

Conclusions

The need for staff to continually develop their skills and knowledge is central to the effective delivery of social care services. As frontline manager you are in a position to play a crucial role in promoting and supporting work-based learning opportunities. Your own approach to learning can be influential in creating a wider commitment to learning in your team. There is a range of work-based learning methods that you can use to develop individuals and teams.

References

Balloch S, McLean J & Fisher M (1999) *Social Services: Working under pressure*. Bristol: Policy Press.

Braye S, Preston-Shoot M & Thorpe A (2006) *Social Work Law in Practice: A research project to explore social work students' law learning on practice placements*. Southampton: SWAP (Social Work and Social Policy Subject Centre).

Chartered Institute of Personnel and Development (2004) *Training and Development, Survey Report*. London: CIPD. Available from: www.cipd.co.uk.

Coulshed V & Mullender A (2006) *Management in Social Work* (third edition). Basingstoke: Palgrave MacMillan.

Darvill G (1997) *The Management of Work-Based Learning: A guide for managers of social care and social work on raising standards of practice*. London: The Stationery Office.

Eraut M (1998) Managers hold the key to developing knowledge and skills. *Professional Manager* **7** (2) 41.

Hudson M (2002) *Managing Without Profit. The art of managing third sector organizations*. London: Directory of Social Change.

Hutchinson S & Purcell J (2003) *Bringing Policies to Life: The vital role of front line managers in people management*. London: CIPD.

Peel M (2003) Managing professional development. In: Seden J & Reynolds J (Eds) (2003) *Managing Care in Practice.* London: Routledge and the Open University.

Smale G (1998) *Managing Change through Innovation*. London: The Stationery Office.

Section 4

Chapter 20

Managing at the Front Line: Managing across professional and agency boundaries

Key points

- **Working in partnership is increasingly seen as the goal for social care services with the intention being to overcome the fragmentation of services so that service users experience a seamless service between different providers.**
- **Research has identified a wide range of factors that influence the success of joint working arrangements and why similar problems have undermined previous attempts at working together.**
- **Managing teams working in partnership means understanding what each profession can contribute to a service and avoiding over-defensive behaviour and unwillingness to explore similarities and differences.**
- **How a frontline manager works with staff, and the relationships and support that are provided can influence the quality of joint working.**

Key words

Pressure for more joint working; service user benefits; multi-agency partnerships; different professional perspectives; openness and willingness to work with others.

An increasing number of frontline managers find themselves responsible for managing services where professionals and support staff from a range of agencies are working together. These multi-agency partnerships are a rapidly growing feature of all services and managers need to understand some of the complex issues involved in the delivery of integrated services at the front line.

There is a long history of agencies co-operating in order to deliver more effective services. But, in spite of exhortations over many years, attempts to achieve more integrated service delivery have often founded as a result of structural, cultural and professional barriers. What has now changed is the Labour government's modernisation agenda with a series of policy initiatives since 1997 that require services to work in partnerships to develop new ways of delivering benefits to service users. Recent examples include, *Our Health, Our Care, Our*

Say (DH, 2006) and *Putting People First* (DH, 2008), which signal a clear intent that the government wants to see more integrated provision for service users.

Services can be seen moving through stages from separatism, where each agency and profession planned and delivered its own contribution in isolation from the contribution of others, through a period of competition where services were placed in competitive relationships to each other, to the current emphasis on partnership where agencies and professionals participate in specific collaborative relationships, driven by strategies that lead in turn to integrated processes at the front line of delivery (Hudson, 2005).

What benefits service users

The drive for more integration of services grows out of the need to improve the service user experience and outcomes. The Care Services Improvement Partnership describes what integration can mean for the service user, including eliminating the fragmentation in service systems, simplifying access and decision-making and reducing the delays and errors that are features in many services. Where integration works best, service users experience services as seamless and the boundaries between organisations are not apparent to them, and increasingly have opportunities for controlling events themselves. All these factors are seen as essential features of effective systems of social care and are difficult to deliver without a commitment to partnership and integration from the political, strategic and operational level of services (CSIP, 2007).

Understanding terminology

One of the problems of working in partnership with other agencies and professionals is the wide range of terms used to describe different service arrangements. It is helpful to clarify exactly what different terms mean as this can help our understanding of what is expected of staff when working in partnership with other services. The following terms tend to be used interchangeably.

- **Integrated service** – refers to a single team of professionals from a range of services working in partnership with other agencies and services.
- **Collaboration** – where there are relations between different agencies and organisations with professionals representing a range of care sectors and groups.
- **Multidisciplinary working** – where groups of staff from different professional backgrounds are brought together to contribute their different areas of knowledge to the same service users.
- **Multi-agency** – a range of agencies working together having shared aims in the delivery of health and social care services.
- **Partnerships** – organisational and individual relationships involving mutual commitments and obligations.

Examples of partnership working

Multi-agency partnership working

A recent example is the multi-agency partnerships required following the introduction of *Safeguarding Adults* (ADSS, 2005), which is intended to secure the sharing of information in cases of adult abuse, and the *National Service Framework for Children and Young People* (DES & DH, 2004), which emphasises the importance of partnership work with local authorities, parents, carers and children, with collaborative working between local authority services and partner agencies.

Housing and support for people with learning disabilities

An example is where two organisations work in partnership to provide different elements of a service, with one organisation providing and managing buildings and the other organisation providing personal care and support to service users. This has grown rapidly with the development of supported living where social care providers are working in partnership with housing associations to provide accommodation for people with special needs.

> *'It has made life a lot easier not having to worry about the bricks and mortar side of things. We can concentrate on providing support to the service users knowing that our partners are responsible for the maintenance of the buildings. We meet regularly with our housing association partners to sort out issues and it's working very well.'*
>
> Service manager, learning disabilities service

Services for older people living in their own homes

To support vulnerable older people who wish to remain in their own homes, community nurses and occupational therapists are working in partnership with local authority social workers to support service users, including flexibility in funding arrangements to enable nurse managers to access local authority budgets to arrange the provision of domiciliary care.

Sure Start

In social care a children's centre manager, for example in a Sure Start project, will combine leadership and team roles working with local communities to determine priorities for children from an early age, as well as working with other professionals and partners to meet strategic performance objectives in relation to educational, social and healthcare needs of children (McCray, 2008).

What contributes to successful joint working?

In a practical guide to integrated working CSIP (2007) has drawn on research undertaken by Bristol University (Cameron *et al*, 2000) which identified some of the factors that are associated with successful joint working. This research also identified why, over many years, the same problems appeared to undermine effective joint working. Among a wide range of factors, the following list of statements are associated with successful joint working:

- senior managers and professional leaders are supportive as good leadership is critical to success
- overall objectives are clear and realistic as this reduces the risk of failure
- resources, including staff skills and time, are adequate to the task
- the negative impact of constant change is minimised as organisational instability can undermine relationships, teamwork and planning
- clashes of professional philosophies, language and tribalism are minimised, with a focus on shared values and collective trust
- the roles and responsibilities of staff are clear and understood as uncertainty is a common cause of failure in joint working
- communication in and between teams and units is good at all levels – failure of communication has been shown to be at the heart of many problems
- staff feel they have 'ownership' of the joint developments
- management accountability is clear and professional support routines are in place
- joint training is provided and team building supported, as both can help overcome deeply set misunderstandings and prejudices at all levels including management and practitioners.

What working in partnership means

The overriding argument for partnership working is that it should provide better outcomes for the service user. This can get lost in the preoccupation with structural issues focusing on the means of delivery, rather than the ends concerned with the quality of life from the perspective of the service user. This latter focus can come from a non-defensive approach by professionals as to who is the best person to assist in achieving this outcome at any one time, using the skill mix of different disciplines positively. It is also about the willingness to align professional expertise and sensitivities for the good of the service rather than the narrow dictates of any one profession (Gilbert, 2003).

Where multidisciplinary working breaks down it is usually due to inadequate leadership and professional insecurity, which leads to professional defensiveness, but where this works well it can offer a rich mix of skills that give service users and their carers more choice and access to a range of expertise. An important basis for multidisciplinary working is staff feeling confident about their own professional knowledge and respecting the knowledge base of other colleagues (Clatworthy, 1999).

Working together can progress through a range of stages, which have been described by Hudson *et al* (1998). These range from *encounters* of an informal ad hoc type, *communication* with formal joint working, *collaboration* with high levels of trust and joint planning and service delivery to *integration* where organisations integrate teams or even merge, with a loss of organisational identity. Most activity will be focused around communication and collaboration, although the development of community trusts, which have merged social services and health service functions are increasingly common.

To work effectively, partnership services have to overcome a number of deeply entrenched barriers including the following factors identified by Hardy *et al* (1992):

- **structural:** where services have been fragmented both across and within agency boundaries
- **procedural:** such as different planning and budget cycles
- **financial:** different funding mechanisms and resource allocation
- **professional:** differences in ideologies, values and professional interests
- **perceived threats:** to status, autonomy and legitimacy.

This list of factors will be familiar to most frontline managers who have worked across the boundaries of different agencies, or striven to deliver more 'joined-up' services in collaboration with colleagues from other agencies.

In an attempt to identify the principles that can assist in creating genuine partnership working, Gorman (1998) has identified a number of important factors that help to provide more effective services for users.

- Reciprocity being at the heart of all partnerships. Is there something in it for both partners?
- Successful partnerships have someone who is willing to act as a convenor and is respected across the agencies involved and is sensitive to the issues that affect the individual partners.
- Partnerships have to work at all levels. Agreements on partnership working are often negotiated at middle management level, but have to be agreed by senior managers and understood by frontline managers and staff who have to implement them.
- Successful partnerships tackle the differences between them, understanding the values and approaches that can be in conflict, bringing them into the open and working with them, and then managing these differences to ensure effective working.
- Focus on outcomes, as in the end effective partnerships should be about better outcomes for service users.

What does this mean for the frontline manager?

Managing in multi-agency or multidisciplinary settings places increasing demands on frontline managers, including: understanding the different value bases between the disciplines, an ability to cope with the variety of professional languages and negotiating the sharing of tasks (Learner & Statham, 2005).

Achieving effective partnership working needs an understanding of the ability to share values, respect and work with cultural differences and agreement of mechanisms of control, reporting and performance measurement, which are dependent on the partner's ability to trust each other. It also requires a management style that includes the ability to work collaboratively, sharing knowledge and skills and devising ways of working that are non-hierarchical and encourage open communication between different groups of staff.

If you are experienced in working with other organisations you will recognise that there can be significant cultural differences between organisations and that the motivation, values and styles of management in the NHS or voluntary sector will be significantly different from local government or the private sector. Each will bring its own cultural assumptions, perspectives, language and preferred ways of working that need to be recognised and managed sensitively.

Implementing partnership working at the front line

Although the development of formal arrangements for partnerships are taken at senior level, it is frontline managers who will be tasked with translating these policies into practice. Ideally, frontline managers will have been involved, at least in a consultative capacity, in the development of these new arrangements, particularly in view of the fact that the importance of performance is critically dependent on the quality of management and practice at the front line (Learner & Statham, 2005). The reality is often different with strategies that failed to take into account issues at the stage of implementation which have implications for senior managers and the introduction of new policies.

Where implementation is well planned and staff are fully involved in working through the detailed issues of working collaboratively with other professionals, there are still issues that need consideration. These require an understanding of the different value bases, the perspectives of different agencies, the culture of different organisations and disciplines, and the ability to cope with different professional languages.

> *'The government's agenda is crystal clear, you will work together by pooling resources to provide better benefits for the user, but it's not easy for managers. You've got to recognise the different cultures, who holds power and how decisions get made. The value base can be quite different and it takes time to get used to the language, the acronyms and just the different way services are provided.'*
>
> Frontline manager, adult services

What can be done to improve working together?

What does working collaboratively involve? McCray (2007) has spelt out some of the key elements to achieving more effective collaboration that can assist the frontline manager, including:

- getting to know other professionals and their roles and how these roles may be evolving in the new context
- thinking about language and terms used and their relevance and meaning to other professionals
- exploring your own assumptions and prejudices about other professional groups and their models of practice
- not making assumptions about values and shared belief of views
- being confident in your own professional practice and ability in collaborative working
- being credible, delivering as promised to action plans and keeping others involved of progress
- gaining consensus on leadership and accountability in specific practice situations.

There are a range of activities that frontline managers can engage in that can increase the effectiveness of working in partnership with other professions. Charlesworth (2003) suggests that team members could give brief talks about their organisation and their roles and responsibilities to increase understanding. Away days can be arranged where staff can learn more about each others' roles and their expectations for collaboration. Above all there is the need to maintain strong leadership and a vision in order that there is a clear focus on the aims of the partnership, including encouraging a range of individuals to assume leadership roles. Particularly important is building trust where there has been no history of partnership working or where there is a legacy of poor relations or there is stereotyping of other professionals that creates a barrier to effective partnership working.

Frontline managers also need to support those staff who find working in partnership with other professionals and agencies threatening and need to understand how the impact of change affects individuals and what processes can be used to enable staff to adapt to working in more complex arrangements.

It should be remembered that many professionals are trained to work autonomously and value their independence. Learning to work as part of a team is not easy and requires practitioners to reorient themselves to working jointly with other professions. These attitudes are primarily inculcated during initial professional training and once qualified staff expect to work autonomously. This means that the education of practitioners needs to reflect the growing arrangements for joint working. Staff need an opportunity through in-service training to reorient themselves to working with others (Peck *et al*, 2002).

Frontline managers' experience

My own research (Scragg, 2006) with frontline managers in a partnership trust highlighted some of the complexities facing them as they led multidisciplinary teams. They were positive about their new role and experienced considerable satisfaction managing in a multidisciplinary context with opportunities to gain a wider professional perspective. The most challenging part of the role was establishing relationships with a range of professionals where role boundaries were still being defined (and in some cases defended).

The most successful managers had developed an understanding of the unique contribution of each team member, with sensitivity towards those members of the team who were still defensive about the new working arrangements. It was the appreciation of what each member of staff could contribute from their own professional expertise that had the potential to enhance partnership working.

One of the factors that was critical in leading a multidisciplinary team was the development of positive relationships with team members. This finding confirmed the view that the personality of a team manager is significant, rather than professional background, as a key factor in their effectiveness and their ability to develop positive relationships with a range of team members from different disciplines (Peck *et al,* 2002).

Conclusions

The thrust of government policy is increasingly towards greater integration of services overcoming some of the barriers and fragmentation that have undermined effective service delivery in the past. Working in partnership with other professionals means recognising the different factors that influence staff and their attitudes to working together. An understanding of the values and beliefs of different professions can help the frontline manager work more sensitively with staff in multidisciplinary or multi-agency settings. Frontline managers also have an important leadership role in creating the conditions where staff from different professions can work together to provide more effective services.

References

Association of Directors of Social Services (2005) *Safeguarding Adults: A national framework of standards for good practice and outcomes in adult protection work.* London: ADSS.

Cameron A, Lart R, Harrison L, Macdonald G & Smith R (2000) *Factors Promoting and Obstacles Hindering Joint Working: A systematic review*. Bristol: Policy Press.

Charlesworth J (2003) Managing across professional and agency boundaries. In: J Seden & J Reynolds (Eds) *Managing Care in Practice*. London: Routledge.

Clatworthy W (1999) Collaborative working. *Journal of Community Nursing* **13** (2) 4–8.

CSIP (2007) *Bringing the NHS and Local Government Together. A practical guide to integrated working*. London: CSIP.

Department of Health (2006) *Our Health, Our Care, Our Say*. London: DH.

Department of Health (2008) *Putting People First*. London: DH.

Department of Education and Skills and Department of Health (2004) *National Service Framework for Children and Young People and Maternity Services*. London: Department of Health.

Gilbert P (2003) *The Value of Everything. Social work and its importance in the field of mental health*. Lyme Regis: Russell House Publishing.

Gorman P (1998) *Managing Multidisciplinary Teams in the NHS*. London: Kogan Page.

Hardy B, Turrell A & Wistow G (1992) *Innovations in Community Care Management.* Aldershot: Avebury.

Hudson B, Exworthy M & Peckham S (1998) *The Integration of Localised and Collaborative Purchasing: A review of the literature and a framework for analysis*. Leeds: Nuffield Institute for Health.

Hudson B (2005) Partnership working and the children's services agenda: is it feasible? *Journal of Integrated Care* **13** (2) 7–12.

Learner E & Statham D (2005) The role of the first line manager: new challenges. *Management Issues in Social Care*, Winter.

McCray J (2007) Reflective practice for collaborative working. In: C Knott & T Scragg (Eds) *Reflective Practice in Social Work*. Exeter: Learning Matters.

McCray J (2008) (Ed) *Nursing and Multi-professional Practice*. London: Sage.

Peck E, Gulliver P & Towell D (2002) *Modernising Partnerships: An evaluation of Somerset's innovations in commissioning and organisation of mental health services – Final report.* London: Institute of Applied Health and Social Policy.

Scragg T (2006) An evaluation of integrated team management. *Journal of Integrated Care* **14** (3) 39–48.

Managing at the Front Line: Greater personalisation of services

Key points

- **The policy of personalisation is intended to give service users greater choice and control over the services they receive.**
- **It has the potential to change service users' relationship with a service as they assume greater responsibility for their own care arrangements.**
- **It involves listening to service users, providing them with information and responding flexibly in planning support, ensuring that they remain in control.**
- **Personalisation has significant implications for services that will need to respond to the demands of service users for non-traditional forms of provision that more closely meet their needs.**

Key words

Service users' choices; control; personal budgets; service change.

The government's vision for social care services is one where service users are able to take greater control of their lives and increasingly assess their own needs and make decisions about how those needs are to be met (DH, 2006; 2007). This is part of the government's strategy to achieve greater user involvement, ranging from consultation on new service initiatives to involving service users in designing and arranging their own services, with the support of professionals where necessary.

Greater service user choice and control has become a guiding principle for services, whether it is concerned with person-centred planning in learning disabilities services, consultation with older people, or support for service users developing their own services. In spite of government intentions and increased sensitivity to the voice of service users there remain considerable barriers to achieving genuine involvement due to professional resistance, power imbalances and inflexible service models, with a significant gap between the rhetoric of service user involvement and the reality of their experience of services (DEMOS, 2008).

In spite of widespread consultation and greater engagement with service users, the basic social care model remains unchanged, with professionals deciding what people need from a fixed menu of services, whether delivered in-house or through the independent sector. This is now slowly beginning to change with the development of more personalised services driven by individual budgets.

A major element in giving service users greater choice and control can be found in the expansion of direct payments and individual budgets, embracing the notion that the needs of individuals should be central to the design of services and that greater choice and control will eventually lead to a range of services that more closely meet what service users want rather than what professionals think are right for people. This signposts an entirely new approach to supporting people who will have the right to make decisions about their own support and the help that can be provided to enable them to live ordinary lives.

This vision for social care services is intended to inform people what they are entitled to and to give them control over their own budgets for services they require. Service users will be in a position to commission their own services and may need support to manage their budget. For service users who do not wish or are unable to manage their own budget and care arrangements, these will be provided by care managers working in a brokerage role or by their family or friends or an independent care broker, irrespective of who provides services. The personalisation of services will have an enormous impact over the next decade. Local authorities will still have a duty of care in that a person's assessed needs are met, but services designed to meet these needs will no longer necessarily be those provided by the local authority directly in contract to the independent sector, but include services and supports chosen by the service users that are more closely tailored to their individual needs.

What users want from services

An important step in understanding what service users want from services is to listen to their voice as this can provide balance in a situation where needs are professionally identified and that may not accord with service users' views of their needs. There is now increasing evidence from service users about their views of services and what they value. Shaping Our Lives (www.shapingourlives.org.uk) is a user-led organisation that has carried out its own research and provides a valuable insight into how people who use services actually view them (Beresford, 2005).

Services are often seen as patchy, inflexible and lack reliability, with staff of variable quality and the employment of agency staff a particular concern. The workforce is seen to operate with poor pay levels and conditions. Direct payments are valued, but currently are an option for very few people, with variable levels of implementation across social services departments and reluctance to extend payments to some user groups such as people with mental health problems.

There is still a need for a local infrastructure of service user organisations to support people. Service users are also concerned about their involvement in consultations and what these actually achieve. They were sceptical about 'user involvement', and experienced 'involvement fatigue' due to too many initiatives. They saw consultation as essentially tokenistic at best (www.shapingourlives.org.uk).

What are service users looking for?

Surveys of what service users are looking for from social care services identified that first and foremost they wanted a service that was easy to navigate, with high-quality and accessible information available at all stages. They valued continuity of care with services joined up. They wished to be equal partners in care, treated with respect and with their family involved. Service users wanted:

- services that are more responsive both in speed and convenience of delivery
- improving assessment and review procedures and flexiblity so they fit in with people's individual circumstances
- services tailored more to the needs and lifestyles of users
- services that are of good quality, with staff having the appropriate skills
- consultation of users and ensuring that their views shape the future of services
- the assurance of stronger safeguards for users through new developments in inspection and complaints procedures
- services should be integrated with staff dealing with boundary problems rather than users. (Farrell *et al*, 1999)

The move towards greater user control

Direct payments were introduced in 1997, following a long campaign by disability groups who wished to control their own budgets and commission their own care through employing their own staff (although restricted by over-regulation). We are now seeing the beginnings of a genuine attempt to transfer power to service users with the piloting of individual budgets in social care services.

Individual budgets are the allocation of funds to a person needing social care, which can be taken as cash or in the form of services, or a mix of the two. It should be the equivalent to the value of the care package that the person would receive through conventional social care services, combined with any other grants and allowances.

The assessment process should be streamlined across agencies and include the option for self-assessment. Individuals can draw up their own support plan for activities services and/or equipment and this is signed off by a social worker or care manager as being capable of meeting the assessed needs. Individuals should have access to brokerage and/or advocacy to help them draw up their plans and manage their budget. Individual budgets have been piloted in 13 local authorities and are currently being evaluated. It is the government's intention to roll out individual budgets to all local authorities. The development of direct payments and individual budgets give much greater control to the individual, opening up the range and availability of services to match needs.

Recent developments

In 2003 a partnership between the Department of Health Valuing People team, Mencap and a group of local authorities came together to help local authorities adopt a self-directed support system for social care in which people with learning difficulties would control how they lived and the support that they needed. Six local authorities originally took part, but this rapidly spread and by 2007, with over 100 local authorities were members of in Control and 2,300 people were receiving personal budgets using this model (DEMOS, 2008).

In addition to the in Control model, there has been a growth in individual budgets that has now spread to over 25,000 providers in the public and voluntary sectors serving approximately 1.7 million people, with people over 65 the largest group of recipients. At the end of 2006 approximately 43,000 people were receiving direct payments and 2,300 were receiving personal budgets for their social care. These arrangements are likely to expand rapidly in the future and it is predicted that within five years several hundred thousand people could be commissioning their social care this way (DEMOS, 2008).

Working with service users and carers and their representatives

This area of development is seen as central to government policy with the government challenging local authorities to engage with users and carers in making services more responsive to their needs. These are priorities for services and it is the government's intention that this is achieved through a range of initiatives.

The DEMOS report *Making it Personal* (2008) describes a number of advantages in the move towards more personalised services, including the following.

- Providing service users with the opportunity to move from being passive recipients to participants in planning and commissioning their own support services.
- A changing relationship with professionals as services change from traditional models, where decisions are made for the service user, to one where professionals become advisers, guides and brokers, providing guidance for service users who make choices for themselves.
- Greater service user involvement creating new ideas, knowledge and information which help shape future services. Bringing more detailed knowledge from service users, their families and friends about what is important and what can be done to support the person, rather than relying on professionals working under pressure designing service packages where service users have little say in the process.
- Service providers needing to adjust to user demands, shifting from inflexible, centralised buildings-based services commissioning through block contracts to more flexible, informal and decentralised provision organised around people's homes, and that allow service users to create new solutions to their needs.

- A shift in power towards service users and away from professionals who retain a critical overview of service quality and outcomes. This means more responsibility for service users in assessing and managing risk along with statutory responsibilities when they become employers.

Potential risks

Although the greater personalisation of services and the provision of individual budgets offer service users greater choice and control, there are also risks associated with these new developments, in terms of funding available for individual support packages and the infrastructure of services available to meet individual need. There will also be issues about those service users who are unable to manage their own services and also those who are vulnerable to the influence of others where safeguards will need to be put in place. These new arrangements also risk the de-professionalisation of services as service users reject established services in favour of more informal and personally arranged support (Managing New Realities, 2008). To meet this potential demand means a radical change in the social care workforce, with the need for a new workforce that will challenge existing professional roles (Skills for Care, 2006).

Implications for social work

Evidence for the DH individual budget pilot programme and in Control suggests that social workers undertaking the care management role have embraced the new approach, as this has provided an opportunity to work within a service context where the values of social work can support practice. These changes have implications for the care management role where the current system has not delivered the benefits that many service users want. It is a system that is constrained by the financial and structural arrangements in local government which do not allow care managers to design services around people's needs as they are often restricted to locating services within existing block contracts, or in-house services, and rationing services as a consequence of the need to work within 'appropriate budgets' (Duffy, 2007). This is where the so-called needs-led assessment turned into a service-led assessment, with all the restrictions that follow from a menu of services that are re-determined rather than designed around the person's needs, and a cause of considerable frustration to social workers who are powerless to create innovative solutions to users' needs.

Frontline managers will also need to support their team members as their practice is questioned by the development of personalisation. Patterns of practice that have developed over many years will need to be challenged (and will also be challenged increasingly by service users who see professionals forcing them into the straitjacket of traditional services that do not meet their needs) and greater trust placed in service users, their families and carers who the pilots have demonstrated are often more sensitive to the sorts of services and other supports that would work for them.

This new policy drive will alter the balance of power in the relationship between services and users because services will no longer have the monopoly on provision as principle provider. That need will be first determined by the user themselves and, through a more transparent process, local authorities will still have a duty of care and will meet that obligation by ensuring that the service user controls their budget and in turn the services they purchase.

This will be particularly so in the case of those service users who have traditionally been seen as less capable of arranging and managing their own lives. This will include older people and disabled people, where current services have fostered dependency unnecessarily and where it has been difficult to envisage that individuals have the capacity to define their own needs and the sorts of services that would meet their needs.

What are the implications for services of these changes?

This shift in policy contains many challenges for managers and professionals. For those social workers who have traditionally undertaken a care management role in assessing service users' needs and arranged services, a shift in responsibility to the service user (and importantly the shift in power) will change their relationship, which has been one of professional recipient. In the past, no matter how much social workers wished to work as equals with service users, they were constrained by local authority systems designed to achieve organisational goals rather than meet the needs of service users.

For people working in local authority adult services, the greatest change will be in the roles of social workers whose work will change as they take on a co-ordinating role in support of service users.

It will be important that care managers and other staff working with service users encourage new ways of thinking about their care, rather than promoting any particular package of care based on existing services or those already purchased by local authorities. It is likely that social workers will become frustrated if the support plans that have the approval of the service user, are held up or declined by budget holding managers who are resistant to signing it off as some elements of the plan are not viewed as social care (Managing New Realities, 2008).

Care managers will also need to become more familiar with a wider range of potential service provision and informal and neighbourhood care than is currently the case. Care managers who have been constrained by the current system in advising the use of a restricted group of mainly institutional services will need to explore a much wider range of options. These may be informal and unfamiliar and in many instances the service user will be best placed to advise on what is most appropriate to meet their needs, but which in the past has not been accessible due to the financial and structural constraints on local authorities.

Implications for provider services

All provider services will need to examine what they provide and whether it meets the needs of service users who will become increasingly sophisticated about what services they wish to purchase. Managers working in provider services will be in a strong position to redesign their services in ways that meet people's needs rather than having to conform to contractual requirements even when these restricted the range of options open to service users.

> *'Increasingly we need to be more flexible and innovative in the way we work, trying to find ways of combining packages of support based on individualised budgets with the more traditional forms of funding such as block contracts. This not only requires us to work more creatively, but also a commitment from service users, their families and commissioners to see the world in a different way that puts the service user firmly in control.'*
>
> Frontline manager, voluntary organisation

As a manager of a provider service you need to understand what service users want from services. For those working in group care the changes will be slower for these services, but increasingly you will see more and more service users or their advisers or brokers making increasing demands on your services to tailor what you can offer to what users want, rather than offering a fixed menu of services. Similarly, staff working for providers will begin to see changes as group care services respond to the demands of service users for more personalised solutions to their needs. As self-directed services grow service users are increasingly likely to choose alternative care arrangements that are no longer dependent on traditional services and the demand for these will decline, with more flexible, personalised services based around people's homes and communities likely to expand. Personal budgets will also bring important changes to the social care workforce as service users purchase more individualised services, with a growth in personal assistants, brokers and advisers (Skills for Care, 2006). Some of these roles will be carried out by family members and friends and have a further affect on the workforce.

Some of the most dramatic changes will need to take place in provider services where traditional patterns of service provision will increasingly need to change as service users gain control of their own funding and make choices that challenge services that foster dependency and squeeze users into regimes that undermine their ability to manage their own affairs and so that services can be designed around their particular needs. Frontline managers will be in the forefront of these changes as services strive to redesign their provision in line with the myriad demands that will be made on them, as service users gain increasing control and power over providers for the first time.

As a frontline manager you will need to be alert to these changes in the long term as you manage the transition. This change will be gradual with changing demand providing services with an opportunity to adjust to new service forms that serve the needs of users for more personalised and individualised services. You will need to start providing more tailored services to individual service users rather than relying on large-scale centralised provision.

If you are managing group services that are delivering contractual services for local authorities, the development of individual budgets could have a significant impact on your services in the long term. Service users with their own budgets will make decisions about packages of care, which may mean that you have to change and adapt your services to meet the needs of service users who no longer wish to use services currently provided. This is already happening in the service where I am a trustee with our day service having to provide activities that are more tailored to the needs and demands of service users who no longer wish to use the traditional service purchased by the local authority on a block contract.

The change to this new agenda will not happen overnight and there will be services where local authorities are locked into long-term block contracts that will make it more difficult for service users to take charge. It is nevertheless important to stress that some traditional provision will still be needed and that some service users will still wish to access this provision in the future. Some service users will not be able to, or want to, control their own budgets and will need others to manage this for them.

These changes will have a significant impact on the management of social care services. They will inevitably progress at a different pace in different localities dependent on local circumstances, but the thrust of policy direction is clear. As the ADSS have argued, there is a continuing pattern of purchasing established services such as residential and nursing care over new and innovative forms of provision despite the policy to seek less institutionalised forms of service provision that more closely match people's needs (ADSS, 2005).

As a frontline manager you are in an ideal position to influence the development of this policy at an implementation level in supporting service users who wish to control their own budgets and make their own arrangements for care; and to support workers who will need to adapt their way of working to support service users, rather than offering a service based on professional views about what people need. This could all lead to an exciting future where services can be tailored more effectively to what people actually want.

Services are now on the cusp of radical change as the shift from a dependency model of care to one based on entitlement with greater choice and control by the service user is being developed across all social care services. Although there are historical factors in terms of organisational policies, systems and processes, the direction of travel is clear and is embedded in government policies that will witness increasing numbers of service users controlling their own budgets and making decisions about the care they receive in the future.

Conclusions

The growth of personalisation with the goal of giving service users more choice and control over the services they receive is now part of government policy and is being developed in all local authorities, although the pace of implementation is varied. The implications of this change for the provision of social care have major consequences for both professionals and services as they respond to greater service user control of their care arrangements.

References

ADSS (2005) *Pressures on Learning Disabilities Services*. London: Association of Directors of Social Services.

Beresford P (2005) 'What services users want' *The Guardian* 23 March.

Department of Health (2006) *Our Health, Our Care, Our Say: A new direction for community services*. London: Department of Health.

Department of Health (2007) *Putting People First: A shared vision and commitment to the transformation of adult social care*. London: Department of Health.

DEMOS (2008) *Making It Personal.* London: DEMOS.

Duffy (2007) Care management and self-directed support. *Journal of Integrated Care* **15** (5) 3–14.

Farrell C, Robinson J & Fletcher P (1999) *A New Era for Community Care? What people want from health, housing and social care services*. London: Kings Fund.

Managing New Realities (2008) Making it personal – individual services in health, housing and social care conference. Information available from: www.societyguardian.co.uk/managingnewrealities.

Skills for Care (2006) *New Type of Worker project 2003–2006*. London: Skills for Care.